W9-ASC-460

Bob Miller's

Math Prep for the ACCUPLACER® Test

Bob Miller

NANUET PUBLIC LIBRARY
149 CHURCH STREET
NANUET, NEW YORK. 10954

Research & Education Association
Visit our website at: www.rea.com

Research & Education Association
61 Ethel Road West
Piscataway, New Jersey 08854
E-mail: info@rea.com

BOB MILLER'S
Math Prep for the ACCUPLACER® Test, 2nd Edition

Published 2017
Copyright © 2016 by Research & Education Association, Inc. Prior edition copyright © 2010 by Research & Education Association, Inc. All rights reserved. No part of this book may be reproduced in any form without permission of the publisher.

Printed in the United States of America

Library of Congress Control Number 2016935858

ISBN-13: 978-0-7386-1211-9
ISBN-10: 0-7386-1211-1

LIMIT OF LIABILITY/DISCLAIMER OF WARRANTY: Publication of this work is for the purpose of test preparation and related use and subjects as set forth herein. While every effort has been made to achieve a work of high quality, neither Research & Education Association, Inc., nor the authors and other contributors of this work guarantee the accuracy or completeness of or assume any liability in connection with the information and opinions contained herein and in REA's software and/or online materials. REA and the authors and other contributors shall in no event be liable for any personal injury, property or other damages of any nature whatsoever, whether special, indirect, consequential or compensatory, directly or indirectly resulting from the publication, use or reliance upon this work.

ACCUPLACER® is a registered trademark of the College Board, which was not involved in the production of, and does not endorse, this product. All other trademarks cited in this publication are the property of their respective owners.

Cover image: © thinkstock.com/michaeljung

REA® is a registered trademark of
Research & Education Association, Inc.

C17

TABLE OF CONTENTS

Author Acknowledgments vi
Dedication vi
Author Biography vii
Other Books by Bob Miller viii
About REA ix
To the Student: A Must-Read x
About the ACCUPLACER Math Test xi

ACCUPLACER Math Review

CHAPTER 1: The Beginnings 1

CHAPTER 2: Arithmetic We Must Know 13

CHAPTER 3: The Power of Exponents 35

CHAPTER 4: The Most Radical Chapter of All 47

CHAPTER 5: Vital Basics of Algebra 61

CHAPTER 6: Equations and Inequalities 87

CHAPTER 7: Words and Word Problems 107

CHAPTER 8: Two or More Unknowns 121

CHAPTER 9: Points, Lines, and More 131

CHAPTER 10: All About Angles and Triangles 147

CHAPTER 11: Other Two-Dimensional Figures 169

CHAPTER 12: Circles, Circles, Circles 183

CHAPTER 13: All About Three-Dimensional Figures 195

CHAPTER 14: More Algebraic Topics 204

CHAPTER 15: All About Trig 225

CHAPTER 16: Topics That Don't Fit in Anywhere Else 251

CHAPTER 17: 12 Practice Tests 293

Arithmetic Practice Tests

Test 1 293
- Test 1 Answers 297
- Test 1 Answer Key 298

Test 2 299
- Test 2 Answers 303
- Test 2 Answer Key 304

Test 3 305
- Test 3 Answers 309
- Test 3 Answer Key 310

Test 4 311
- Test 4 Answers 315
- Test 4 Answer Key 316

Elementary Algebra Practice Tests

Test 1 317
- Test 1 Answers 319
- Test 1 Answer Key 320

Test 2 321
- Test 2 Answers 323
- Test 2 Answer Key 324

Test 3 325
- Test 3 Answers 328
- Test 3 Answer Key 329

Test 4 330
- Test 4 Answers 333
- Test 4 Answer Key 334

College-Level Mathematics Practice Tests

Test 1 335
Test 1 Answers 341
Test 1 Answer Key 343
Test 2 344
Test 2 Answers 349
Test 2 Answer Key 351
Test 3 352
Test 3 Answers 357
Test 3 Answer Key 359
Test 4 360
Test 4 Answers 365
Test 4 Answer Key 367

INDEX 369

AUTHOR ACKNOWLEDGMENTS

I have many people to thank.

I thank my wife, Marlene, who makes life worth living, who is truly the wind under my wings. I thank the rest of my family: children Sheryl and Eric and their spouses Glenn and Wanda (who are also like my children); grandchildren Kira, Evan, Sean, Sarah, Ethan, and Noah; my brother Jerry; and my parents, Cele and Lee; and my in-law parents, Edith and Siebeth.

I thank Larry Kling and Kelli Wilkins at REA for making the second edition of this book possible.

I thank Martin Levine for making my whole writing career possible. I have been negligent in thanking my great math teachers of the past. I thank Mr. Douglas Heagle, Mr. Alexander Lasaka, Mr. Joseph Joerg, and Ms. Arloeen Griswold, the best math teacher I ever had, of George W. Hewlett High School; Ms. Helen Bowker of Woodmere Junior High; and Professor Pinchus Mendelssohn and Professor George Bachman of Polytechnic University.

The death of Professor Bachman was an extraordinary loss to our country, which produces too few advanced degrees in math. Every year, two or three of Professor Bachman's students would receive a Ph.D. in math, and even more would receive their M.S. in math. In addition, he wrote four books and numerous papers on subjects that had never been written about or had been written so poorly that nobody could understand the material. His teachings and writings were clear and memorable.

As usual, the last three thanks go to three terrific people: a great friend, Gary Pitkofsky; another terrific friend and fellow lecturer, David Schwinger; and my cousin, Keith Robin Ellis, the sharer of our dreams.

Bob Miller

DEDICATION

To my wife, Marlene. I dedicate this book and everything else I ever do to you.
I love you very, very much.

AUTHOR BIOGRAPHY

I received my B.S. in the Unified Honors Program sponsored by the Ford Foundation and my M.S. in math from Polytechnic Institute of NYU. After teaching my first class, as a substitute for a full professor, I heard one student say to another upon leaving the classroom, "At least we have someone who can teach the stuff." I was hooked forever on teaching. Since then, I have taught at CUNY, Westfield State College, Rutgers, and Poly.

No matter how I feel, I always feel a lot better when I teach. I always feel great when students tell me they used to hate math or couldn't do math and now they like it more and can do it better.

My main blessing is my family. I have a fantastic wife in Marlene. My kids are wonderful, and my grandchildren are terrific. My hobbies are golf, bowling, bridge, crossword puzzles, and Sudoku. My ultimate goals are to write a book to help parents teach their kids math, a high school text that will advance our kids' math abilities and a calculus text students can actually understand. To me, teaching is always a great joy. I hope that I can give some of that joy to you.

OTHER BOOKS BY BOB MILLER

Bob Miller's Math for the ACT

Bob Miller's Math for the GMAT

Bob Miller's Math for the GRE

Bob Miller's Math for the TABE

Bob Miller's Basic Math and Pre-Algebra for the Clueless, Second Edition

Bob Miller's Algebra for the Clueless, Second Edition

Bob Miller's Geometry for the Clueless, Second Edition

Bob Miller's SAT Math for the Clueless, Second Edition

Bob Miller's Precalc for the Clueless, Third Edition

Bob Miller's High School Calc for the Clueless

Bob Miller's Calc 1 for the Clueless, Second Edition

Bob Miller's Calc 2 for the Clueless, Second Edition

Bob Miller's Calc 3 for the Clueless

ABOUT REA

Founded in 1959, Research & Education Association (REA) is dedicated to publishing the finest and most effective educational materials—including study guides and test preps—for students of all ages. Today, REA's wide-ranging catalog is a leading resource for students, teachers, and other professionals. Visit *www.rea.com* to see a complete listing of all our titles.

REA ACKNOWLEDGMENTS

In addition to our author, we would like to thank Pam Weston, Publisher, for setting the quality standards for production integrity and managing the publication to completion; Larry B. Kling, Director of Editorial Services, for overall direction; Kelli Wilkins, Editor, for project management; Ellen Gong for proofreading; and Eve Grinnell, Graphic Artist, for designing our cover and typesetting this edition.

Back-cover photo of author by Eric L. Miller.

TO THE STUDENT: A MUST-READ

Congratulations! You are about to begin college, a new and exciting time in your life. This book is written to help you refresh and improve your math skills so you can be properly placed at the right level in your college math courses.

Bob Miller's Math Prep for the ACCUPLACER Test will supply you with the material you need to succeed on the math portion of the ACCUPLACER exam. The ACCUPLACER is a computer-adaptive test (CAT), which means you will be in front of a computer. The computer will ask you a question. Using paper and pencil, you will figure out the answer. You must enter an answer by clicking on a choice on the computer screen.

Because the test is adaptive, your responses determine the questions that are presented to you. A correct answer will give you a more difficult math question, and an incorrect answer will give you an easier one. In this way, no two people take the exact same test. You must answer each question before continuing. If you do not know an answer, try to eliminate one or more incorrect choices, and then make your best guess.

If, after you receive your ACCUPLACER test results, you think you placed too low, most colleges will retest you if you ask. If you feel you placed too high, you do not have to accept the placement. After all, you are paying for your education.

In studying with this book, you may want to take the practice tests more than once. If you miss questions on the practice tests, go back to the review and restudy the appropriate chapter(s). Our answer keys include a chart that will help you zero in on the content you need to review. This second edition of *Bob Miller's Math Prep for the ACCUPLACER Test* provides you with all the tools you need to succeed on this important college placement exam.

Good luck in college and in your future!

Bob Miller

ABOUT THE ACCUPLACER MATH TEST

What is the ACCUPLACER test?

The ACCUPLACER is a battery of computer-adaptive college-placement assessments developed by the College Board. These tests are designed to evaluate your skills in reading, writing, and mathematics as you prepare to enroll in college-level courses. The tests help identify your strengths and needs in each subject area. This book will help you prepare for three of these tests:

- Arithmetic
- College-Level Math
- Elementary Algebra

Over 7.5 million ACCUPLACER tests are administered each year to some 2.5 million students in more than 1,500 high schools, community colleges, four-year colleges, and technical schools around the world.

Your ACCUPLACER test results, along with other information, will be used by academic advisors and counselors to place you in the appropriate courses at your college or technical school.

What is the format of the ACCUPLACER math tests?

The ACCUPLACER math tests are strictly multiple-choice and computer-adaptive. This means that the test questions you get are based on your responses to the questions you've already answered. The test's difficulty level rises or falls depending on your answers as you work your way through the test. The number of questions also varies depending on which ACCUPLACER tests you take. Some tests have 12 questions, while others have up to 40.

ACCUPLACER tests are untimed, so you can move at a comfortable pace. It's important to give yourself enough time to do your best and complete the test because your results are the key factor in determining the college-level courses you can take.

How is the ACCUPLACER scored? What happens if I fail the test?

You cannot fail the ACCUPLACER test, because a placement test can't be passed or failed. Still, you'll want to bear in mind that the ACCUPLACER test score range is between 20 and 120 points. Each school's faculty determines their own cut scores based upon the skill levels they deem necessary for success in a college course. Any questions on scoring should be directed to your academic advisor. Again, these tests are used only for placement purposes, so they're intended to give you and your college an accurate measure of your academic skills.

The test is scored online and individual student score reports are available immediately after testing. Reports deliver student test scores, customized messages from the institution where you're enrolled, and information about course placement based on institutional placement scores. It's a good idea to ask your academic advisor about testing policies and procedures. That way, you're prepared and can plan accordingly.

How should I deal with field-tested questions?

Occasionally, the test developer field-tests new questions to refresh and expand the test-item pool. Our best advice on how to handle these is not to give them a second thought. You will be notified about such questions in the "What's New" feature of the ACCUPLACER testing system. And they will not count toward your score. So just take them in stride.

How do I make an appointment to take the ACCUPLACER test?

Your college's advising and counseling office will have all the information you need to make an appointment for the test. Your academic advisors will be able to answer specific questions you have about testing dates, fees, and other requirements. If your college is located in another state, contact your institution's test center to learn more about remote testing options. With the permission of both institutions, you may be able to take the ACCUPLACER test at a local college. On test day, bring your photo ID and any special materials that your college has requested. Contact your institution's test center for more details.

What can I bring on test day?

It's your responsibility to consult ACCUPLACER registration materials so you know what you can and cannot bring into the test center. This information may be updated by the College Board from time to time. Rest assured, however, that the testing center will provide you with pens, pencils, and scrap paper.

What will I be tested on?

There are three math tests on the ACCUPLACER test: Arithmetic, College-Level Math, and Elementary Algebra. The following is a breakdown of the math topics tested on the ACCUPLACER test.

Arithmetic

The Arithmetic test, comprised of 17 questions, measures your ability to perform basic arithmetic operations and to solve problems that involve fundamental arithmetic concepts. There are three types of Arithmetic questions:

- Operations with whole numbers and fractions: Topics included in this category are addition, subtraction, multiplication, division, recognizing equivalent fractions and mixed numbers, and estimating.
- Operations with decimals and percents: Topics include addition, subtraction, multiplication, and division with decimals. Percent problems, recognition of decimals, fraction and percent equivalencies, and problems involving estimation are also given.
- Applications and problem solving: Topics include rate, percent, and measurement problems, simple geometry problems, and distribution of a quantity into its fractional parts.

College-Level Math

The College-Level Math test, comprised of 20 questions, measures your ability to solve problems that involve college-level mathematics concepts. There are five types of College-Level Math questions:

- Algebraic operations: Topics include simplifying rational algebraic expressions, factoring, expanding polynomials, and manipulating roots and exponents.
- Solutions of equations and inequalities: Topics include the solution of linear and quadratic equations and inequalities, equation systems and other algebraic equations.

- Coordinate geometry: Topics include plane geometry, the coordinate plane, straight lines, conics, sets of points in the plane, and graphs of algebraic functions.
- Applications and other algebra topics: Topics include complex numbers, series and sequences, determinants, permutations and combinations, fractions and word problems.
- Functions and trigonometry: Topics include polynomials, algebraic, exponential, and logarithmic and trigonometric functions.

Elementary Algebra

The Elementary Algebra test, comprised of 12 questions, measures your ability to perform basic algebraic operations and to solve problems involving elementary algebraic concepts. There are three types of Elementary Algebra questions:

- Operations with integers and rational numbers: Topics include computation with integers and negative rationals, the use of absolute values, and ordering.
- Operations with algebraic expressions: Topics include the evaluation of simple formulas and expressions, adding and subtracting monomials and polynomials, multiplying and dividing monomials and polynomials, the evaluation of positive rational roots and exponents, simplifying algebraic fractions, and factoring.
- Solution of equations, inequalities, word problems: Topics include solving linear equations and inequalities, solving quadratic equations by factoring, solving verbal problems presented in an algebraic context, including geometric reasoning and graphing, and the translation of written phrases into algebraic expressions.

Free ACCUPLACER resources

Sharpening your academic skills by answering sample test questions can lead you to improve your score and can help boost your confidence when you take the actual ACCUPLACER test. This exercise is especially helpful if you've been away from school for more than a year.

The College Board offers a free PDF of examples of the types of questions you'll see when you take the ACCUPLACER test. It is useful for some fast and simple preparation prior to taking the test. For more information and to download the PDF, visit: *https://accuplacer.collegeboard.org/students/accuplacer-tests*

In addition, the College Board also provides a Web-Based Study App. You can access hundreds of questions to familiarize yourself with the ACCUPLACER test's basic content and experience. Visit *https://accuplacerpractice.collegeboard.org/login* for more information.

Test-Taking tips

Know the format of the test. Familiarize yourself with the format of the ACCUPLACER test beforehand by logging on to the College Board website at *https://accuplacer.collegeboard.org*. Waiting until test day to learn about the test risks injecting needless anxiety into your testing experience.

Read all the questions … completely. Make sure you understand each question before looking for the right answer. Reread the question if it doesn't make sense.

Read all of the answers to a question. Just because you think you found the correct response right away, do not assume that it's the best answer. The last answer choice might be the correct answer.

Use the process of elimination. Stumped by a question? Don't make a random guess. Eliminate as many of the answer choices as possible. By eliminating answer choices, you give yourself a better chance of getting the item correct, since there will be fewer choices left from which to make your guess.

Be sure that your answer registers. Look at the screen to see that your mouse-click causes the pointer to darken the proper oval. If your answer doesn't register, you won't be able to move on to the next question.

Making the most of your college plan

After you take the ACCUPLACER test, you may want to consider returning to the testing center after exploring your options with the College Board's CLEP credit-by-exam program.

"CLEP" is short for College-Level Examination Program. Some colleges award as many as 60 academic credits through the CLEP exam process, so your savings in tuition and time could be substantial. Visit the College Board at www.collegeboard.org/clep to find out more about the program's 33 exams. Then, come to REA for an extensive catalog of CLEP test preps to help you pass the CLEP exams and accelerate your degree.

CHAPTER 1: *The Beginnings*

"You have a 60-inch ribbon. If you cut it so that each piece is 6 inches, how many cuts do you make?"

Our great adventure begins with our basic terms. It is very important to understand what a question asks as well as how to answer it. The word *numbers* has many meanings, as we start to see here.

NUMBERS

Whole numbers: 0, 1, 2, 3, 4, . . .

Integers: 0, ± 1, ± 2, ± 3, ± 4, . . . , where ± 3 stands for both $+3$ and -3.

Positive integers are integers that are greater than 0. In symbols, $x > 0$, x is an integer.

Negative integers are integers that are less than 0. In symbols, $x < 0$, x is an integer.

Even integers: 0, ± 2, ± 4, ± 6, . . .

Odd integers: ± 1, ± 3, ± 5, ± 7, . . .

Inequalities

For any numbers represented by a, b, c, or d on the number line:

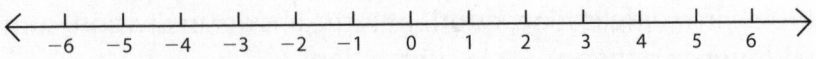

We say $c > d$ (c is greater than d) if c is to the right of d on the number line.

We say $d < c$ (d is less than c) if d is to the left of c on the number line.

$c > d$ is equivalent to $d < c$.

$a \leq b$ means $a < b$ or $a = b$; likewise, $a \geq b$ means $a > b$ or $a = b$.

Example 1: $4 \leq 7$ is true because $4 < 7$; $9 \leq 9$ is true because $9 = 9$; but $7 \leq 2$ is false because $7 > 2$.

Example 2: Find all integers between -4 and 5.

Solution: $\{-3, -2, -1, 0, 1, 2, 3, 4\}$.

Notice that the word *between* does *not* include the endpoints.

Example 3: Graph all the multiples of five between 20 and 40 inclusive.

Solution:

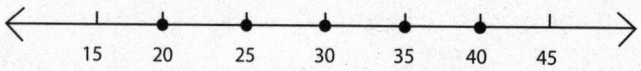

Notice that *inclusive* means to include the endpoints.

Odd and Even Numbers

Here are some facts you should know about odd and even integers.

- The sum of two even integers is even.
- The sum of two odd integers is even.
- The sum of an even integer and an odd integer is odd.
- The product of two even integers is even.
- The product of two odd integers is odd.
- The product of an even integer and an odd integer is even.
- If n is even, n^2 is even. If n^2 is even and n is an integer, then n is even.
- If n is odd, n^2 is odd. If n^2 is odd and n is an integer, then n is odd.

OPERATIONS ON NUMBERS

Product is the answer in multiplication, **quotient** is the answer in division, **sum** is the answer in addition, and **difference** is the answer in subtraction.

Because $3 \times 4 = 12$, 3 and 4 are said to be **factors** or **divisors** of 12, and 12 is both a **multiple** of 3 and a **multiple** of 4.

A **prime** is a positive integer with exactly two distinct factors, itself and 1. The number 1 is not a prime because only $1 \times 1 = 1$. It might be a good idea to memorize the first eight primes:

2, 3, 5, 7, 11, 13, 17, and 19

The number 4 has more than two factors: 1, 2, and 4. Numbers with more than two factors are called **composites**. The number 28 is a **perfect** number because if we add the factors less than 28, they add to 28.

Example 4: Write all the factors of 28.

Solution: 1, 2, 4, 7, 14, and 28.

Example 5: Write 28 as the product of prime factors.

Solution: $28 = 2 \times 2 \times 7$.

Example 6: Find all the primes between 70 and 80.

Solution: 71, 73, 79. How do we find this easily? First, because 2 is the only even prime, we have to check only the odd numbers. Next, we have to know the **divisibility rules**:

- A number is divisible by 2 if it ends in an even number. We don't need this here because then it can't be prime.
- A number is divisible by 3 (or 9) if the sum of the digits is divisible by 3 (or 9). For example, 456 is divisible by 3 because the sum of the digits is 15, which is divisible by 3 (it's not divisible by 9, but that's okay).
- A number is divisible by 4 if the number named by the last two digits is divisible by 4. For example, 3936 is divisible by 4 because 36 is divisible by 4.
- A number is divisible by 5 if the last digit is 0 or 5.
- The rule for 6 is a combination of the rules for 2 and 3.
- It is easier to divide by 7 than to learn the rule for 7.
- A number is divisible by 8 if the number named by the last *three* digits is divisible by 8.
- A number is divisible by 10 if it ends in a zero, as you know.
- A number is divisible by 11 if the difference between the sum of the even-place digits (2nd, 4th, 6th, etc.) and the sum of the odd-place digits (1st, 3rd, 5th, etc.) is a multiple of 11. For example, for the number 928,193,926: the sum of the odd digits (9, 8, 9, 9, and 6) is 41; the sum of the even digits (2, 1, 3, and 2) is 8; and $41 - 8$ is 33, which is divisible by 11. So 928,193,926 is divisible by 11.

That was a long digression!!!!! Let's get back to Example 6.

We have to check only 71, 73, 75, 77, and 79. The number 75 is not a prime because it ends in a 5. The number 77 is not a prime because it is divisible by 7. To see if the other three are prime, for any number less than 100 you have to divide by the primes 2, 3, 5, and 7 only. You will quickly find that 71, 73, and 79 are primes.

Rules for Operations on Numbers

Note *( ) are called parentheses (singular: parenthesis); [ ] are called brackets; { } are called braces.*

Rules for adding signed numbers

1. If all the signs are the same, add the numbers and use that sign.
2. If two signs are different, subtract them, and use the sign of the larger numeral.

Example 7: **a.** $3 + 7 + 2 + 4 = +16$ **c.** $5 - 9 + 11 - 14 = 16 - 23 = -7$

b. $-3 - 5 - 7 - 9 = -24$ **d.** $2 - 6 + 11 - 1 = 13 - 7 = +6$

Rules for multiplying and dividing signed numbers

Look at the minus signs only.

1. Odd number of minus signs—the answer is minus.
2. Even number of minus signs—the answer is plus.

Example 8: $\dfrac{(-4)(-2)(-6)}{(-2)(+3)(-1)} =$

Solution: Five minus signs, so the answer is minus, -8.

Rule for subtracting signed numbers

The sign $(-)$ means subtract. Change the problem to an addition problem.

Example 9: **a.** $(-6) - (-4) = (-6) + (+4) = -2$

b. $(-6) - (+2) = (-6) + (-2) = -8$, because it is now an adding problem.

Order of Operations

In doing a problem such as $4 + 5 \times 6$, the **order of operations** tells us whether to multiply or add first:

1. If given letters, substitute in parentheses the value of each letter.
2. Do operations in parentheses, inside ones first, and then the tops and bottoms of fractions.
3. Do exponents next. (Chapter 3 discusses exponents in more detail.)

4. Do multiplications and divisions, left to right, as they occur.
5. The last step is adding and subtracting. Left to right is usually the safest way.

Example 10: $4 + 5 \times 6 =$

Solution: $4 + 30 = 34$

Example 11: $(4 + 5)6 =$

Solution: $(9)(6) = 54$

Example 12: $1000 \div 2 \times 4 =$

Solution: $(500)(4) = 2000$

Example 13: $1000 \div (2 \times 4) =$

Solution: $1000 \div 8 = 125$

Example 14: $4[3 + 2(5 - 1)] =$

Solution: $4[3 + 2(4)] = 4[3 + 8] = 4(11) = 44$

Example 15: $\dfrac{3^4 - 1^{10}}{4 - 10 \times 2} =$

Solution: $\dfrac{81 - 1}{4 - 20} = \dfrac{80}{-16} = -5$

Example 16: If $x = -3$ and $y = -4$, find the value of:

a. $7 - 5x - x^2$

b. $xy^2 - (xy)^2$

c. $\dfrac{y^2 - x^3 - 3}{4 - 2xy}$

Solutions: **a.** $7 - 5x - x^2 = 7 - 5(-3) - (-3)^2 = 7 + 15 - 9 = 13$

b. $xy^2 - (xy)^2 = (-3)(-4)^2 - ((-3)(-4))^2 = (-3)(16) - (12)^2$

$= -48 - 144 = -192$

c. $\dfrac{y^2 - x^3 - 3}{4 - 2xy} = \dfrac{(-4)^2 - (-3)^3 - 3}{4 - 2(-3)(-4)} = \dfrac{16 + 27 - 3}{4 - 24} = \dfrac{40}{-20} = -2$

Before we get to the exercises, let's talk about ways to describe a group of numbers (data).

DESCRIBING DATA

Four of the measures that describe data are used on the ACCUPLACER. The first three are measures of central tendency; the fourth, the range, measures the span of the data.

Mean: Add up the numbers and divide by how many numbers you have added up.

Median: Middle number. Put the numbers in numeric order and see which one is in the middle. If there are two "middle" numbers, which happens with an even number of data points, take their average.

Mode: Most common numbers—those that appear the most times. A set with two modes is called bimodal. There can actually be any number of modes.

Range: Highest number minus the lowest number.

Example 17: Find the mean, median, mode, and range for 5, 6, 9, 11, 12, 12, and 14.

Solutions: Mean: $\dfrac{5+6+9+11+12+12+14}{7} = \dfrac{69}{7} = 9\dfrac{6}{7}$

Median: 11

Mode: 12

Range: $14 - 5 = 9$

Example 18: Find the mean, median, mode, and range for 4, 4, 7, 10, 20, 20.

Solutions: Mean: $\dfrac{4+4+7+10+20+20}{6} = \dfrac{65}{6} = 10\dfrac{5}{6}$

Median: For an even number of points, it is the mean of the middle two:

$\dfrac{7+10}{2} = 8.5$

Mode: There are two: 4 and 20 (blackbirds?)

Range: $20 - 4 = 16$

Example 19: Jim received grades of 83 and 92 on two tests. What grade must the third test be in order to have an average (mean) of 90?

Solution: There are two solution methods.

Method 1: To get a 90 average on three tests, Jim needs $3(90) = 270$ points. So far, he has $83 + 92 = 175$ points. So Jim needs $270 - 175 = 95$ points on the third test.

Method 2 (my favorite): 83 is -7 from 90. 92 is $+2$ from 90. $-7 + 2 = -5$ from the desired 90 average. Jim needs $90 + 5 = 95$ points on the third test. (Jim needs to "make up" the 5-point deficit, so add it to the average of 90.)

The second method is my choice, but it is always your choice which method you can do easier and faster.

Finally, after a long introduction, we get to some multiple-choice exercises.

Exercise 1: If $x = -5$, the value of $-3 - 4x - x^2$ is

A. -48
B. -8
C. 2
D. 13
E. 4

Exercise 2: $-0(2) - \frac{0}{2} - 2 =$

A. 0
B. -2
C. -4
D. -6
E. Undefined

Exercise 3: The scores on three tests were 90, 91, and 98. What does the score on the fourth test have to be in order to get exactly a 95 average (mean)?

A. 97
B. 98
C. 99
D. 100
E. Not possible

Exercise 4: On a true-false test, 20 students scored 90, and 30 students scored 100. The sum of the mean, median, and mode is

A. 300
B. 296
C. 295
D. 294
E. 275

Exercise 5: On a test, m students received a grade of x, n students received a grade of y, and p students received a grade of z. The average (mean) grade is:

A. $\frac{mxnypz}{xyz}$
B. $\frac{mx + ny + pz}{x + y + z}$
C. $\frac{mx + ny + pz}{xyz}$
D. $\frac{mx + ny + pz}{m + n + p}$
E. $mnp + \frac{xyz}{x + y + z}$

Exercise 6: The largest positive integer in the following list that divides evenly into 2,000,000,000,000,003 is

A. 33
B. 11
C. 10
D. 3
E. 1

For Exercises 7–9, use the following numbers: 8, 10, 10, 16, 16, 18

Exercise 7: The mean is

A. 8
B. 10
C. 13
D. 16
E. There are two of them.

Exercise 8: The median is

A. 8
B. 10
C. 13
D. 16
E. There are two of them.

Exercise 9: The mode is

A. 8
B. 10
C. 13
D. 16
E. There are two of them.

Sometimes statistics are given in frequency distribution tables, such as this one showing the grades Sandy received on 10 English quizzes.

Sandy's Quiz Scores

	Grade	Number
	100	4
	98	3
	95	2
	86	1
Total		10

Use this chart for Exercises 10–12.

Exercise 10: The mean is

A. 96
B. 97
C. 98
D. 99
E. 100

Exercise 11: The median is

A. 96
B. 97
C. 98
D. 99
E. 100

Exercise 12: The mode is

A. 96
B. 97
C. 98
D. 99
E. 100

A Let's look at the answers.

Answer 1: **B:** $-3 - 4(-5) - (-5)^2 = -3 + 20 - 25 = -8$.

Answer 2: **B:** $0 - 0 - 2 = -2$.

Answer 3: **E:** $95(4) = 380$ points; $90 + 91 + 98 = 279$ points. The fourth test would have to be $380 - 279 = 101$.

Answer 4: **B:** The median is 100; the mode is 100; for the mean, we can use 2 and 3 instead of 20 and 30 because the ratio is the same: $\frac{2(90) + 3(100)}{5} = 96$.

So the sum is $100 + 100 + 96 = 296$.

Answer 5: **D:** $Mean = \frac{\text{total value of items}}{\text{total number of items}} = \frac{mx + ny + pz}{m + n + p}$.

Answer 6: **E:** The number is not divisible by 11 because $3 - 2 = 1$, which is not a multiple of 11. It is not divisible by 3 because $3 + 2 = 5$ is not divisible by 3. Because it is not divisible by 11 or 3, it is not divisible by 33. Finally, it is not divisible by 10 because it does not end in a 0. So the answer is E because all numbers are divisible by 1.

Answer 7: **C:** $\frac{8 + 10 + 10 + 16 + 16 + 18}{6} = 13$.

Answer 8: **C:** There are an even number of numbers, so we have to take the average of the middle two: $\frac{10 + 16}{2} = 13$.

Answer 9: **E:** It's bimodal; the modes are 10 and 16, each appearing twice.

Answer 10: **B:** The mean is the longest measure to compute:

$$\frac{4 \times 100 + 3 \times 98 + 2 \times 95 + 86}{10} = 97.$$

Answer 11: **C:** The median is determined by putting all of the numbers in order, so we have 100, 100, 100, 100, 98, 98, 98, 95, 95, 86. The middle terms are 98 and 98, so the median is 98.

Answer 12: **E:** The mode is 100 because that is the most common score; there are four of them.

Chapter 1 Test

1. List the even integers between -4 and 7.
2. List the multiples of seven between 42 and 73 inclusive.
3. List the primes between 100 and 110.
4. Write all the factors of 24.
5. Write 90 as the product of prime factors.
6. Why is 4.56×10^{1000} divisible by 3 and not by 9?
7. $9 - 3^3 - 1^4 =$
8. $7 - 3(-5) - 7(2) =$
9. $-1^{10} + (-1)^{11} =$
10. $\dfrac{(-2)^3 - (-2)^4}{8-(-4)^2} =$

For questions 11 and 12, let $a = -5$ and $d = 3$; evaluate the following:

11. $ad - a^2$
12. $ad^2 - (ad)^2$

For questions 13–16, consider the set {1, 2, 2, 5, 6, 8, 10}.

13. Find the mean.
14. Find the mode.
15. Find the median.
16. Find the range.

For questions 17–20, consider the set $\{-9, -9, -9, 4, 4, 4\}$.

17. Find the mean.
18. Find the mode.
19. Find the median.
20. Find the range.

Answers to Chapter 1 Test

1. $-2, 0, 2, 4$, and 6; "between" doesn't include the end numbers. Remember, 0 is an even integer.
2. 42, 49, 56, 63, and 70; "inclusive" means to include the end number.
3. 101, 103, 107, and 109 (101 and 103, 107 and 109 are sometimes called twin primes because they differ by 2; see if you can find more of them).
4. 1, 2, 3, 4, 6, 8, 12, and 24.
5. $2 \times 3 \times 3 \times 5$, or $2 \times 3^2 \times 5$.
6. $4 + 5 + 6 = 15$ is divisible by 3 but not by 9.
7. $9 - 27 - 1 = -19$.
8. $7 + 15 - 14 = 8$.
9. $-1 + (-1) = -2$.
10. $\dfrac{(-8 - 16)}{(8 - 16)} = \dfrac{-24}{-8} = 3$.
11. $(-5)(3) - (-5)^2 = -15 - 25 = -40$.
12. $(-5)(3)^2 - [(-5)(3)]^2 = -(5)(9) - (-15)^2 = -45 - 225 = -270$.
13. $\dfrac{34}{7} = 4\dfrac{6}{7}$.
14. 2.
15. 5.
16. $10 - 1 = 9$.
17. $-\dfrac{15}{6} = -2\dfrac{1}{2}$.
18. -9 and 4 are both modes (bimodal).
19. $\dfrac{(-9 + 4)}{2} = -2\dfrac{1}{2}$.
20. $4 - (-9) = 13$.

Let's do some decimals, fractions, and percentages.

Answer to "Bob Asks": 9 cuts.

CHAPTER 2: *Arithmetic We Must Know*

"A 60-foot fence is in front of my house.
A fence post is needed every 6 feet.
How many fence posts are needed?"

Although the ACCUPLACER does allow calculators, it is necessary to be able to do some of the work without the calculator, especially fractions. Let's start with decimals.

DECIMALS

Rule 1: When adding or subtracting, line up the decimal points.

Example 1: Add: 3.14 + 234.7 + 86

Solution:

$$\begin{array}{r} 3.14 \\ 234.7 \\ +\ 86. \\ \hline 323.84 \end{array}$$

Example 2: Subtract: 56.7 − 8.82

Solution:

$$\begin{array}{r} 56.70 \\ -\ 8.82 \\ \hline 47.88 \end{array}$$

Rule 2: In multiplying numbers, count the number of decimal places to the right of the decimal point for each number and add them. In the product, this will be the number of decimal places to the right of the decimal point.

Example 3: Multiply 45.67 by .987.

Solution: The answer will be 45.07629. You will need to know the answer has five decimal places to the right.

Example 4: Multiply 2.8 by .6:

Solution: The answer is 1.68.

Example 5: What is the value of $2b^2 - 2.4b - 1.7$ if $b = .7$?

Solution: The standard way to do this type of problem is to directly substitute.
$2(.7)^2 - 2.4(.7) - 1.7 = 2(.49) - 1.68 - 1.7 = .98 - 1.68 - 1.7 = -2.40.$

You will be given answers from which to choose, so approximations may work just as well and are quicker. $(.7)(.7) = .49$; times 2 is .98, or approximately 1. -2.4 times $.7 = 1.68$ or approximately 1.7; $1 - 1.7 = -.7$; added to -1.7 is -2.4. In this case, we get the same numerical value. If we didn't, our approximate answer would still probably be closer to the correct answer than to the other choices.

In the real world, and on this test, a good skill to save time is the ability to make reasonable approximations. This book will point out problems that can be approximated.

Rule 3: When you divide, move the decimal point in the divisor and the dividend the same number of places to the right.

Example 6: Divide 23.1 by .004.

Solution: In our heads, we write the problem as $\frac{23.1}{.004}$. We multiply the numerator (top number) and denominator (bottom number) by 1000 to move the decimal points the same number of places to get whole numbers. We get $\frac{23.1}{.004} = \frac{23.1 \times 1000}{.004 \times 1000} = \frac{23100}{4} = 5775$. Note that when we multiply by 1, the fraction doesn't change value. Since $\frac{1000}{1000} = 1$, multiplying $\frac{23.1}{.004}$ by $\frac{1000}{1000}$ doesn't change its value, but is much easier to perform division.

Rule 4: When reading a number with a decimal, read the whole part, only say the word *and* when you reach the decimal point, then read the part after the decimal point as if it were a whole number, and say the last decimal place. Whew!

Example 7:

Number	Read
4.3	Four and three tenths

2,006.73	Two-thousand six and seventy-three hundredths
1,000,017.009	One million seventeen and nine thousandths

Q Let's do a couple of multiple-choice exercises.

Exercise 1: Which is smallest?

A. .04
B. .0401
C. .04001
D. .04444
E. .041

Exercise 2: $\frac{7}{100} + \frac{8}{10,000} + \frac{9}{1,000,000} =$

A. 789
B. .0789
C. .07089
D. .070809
E. .078009

A Let's look at the answers.

Answer 1: **A:** All have the same tenths (0) and hundredths (4). The rest of the places of A are zeros, so it's the smallest.

Answer 2: **D:** Hundredths is the second decimal place, ten-thousandths is the fourth, and millionths is the sixth place.

Sometimes you have to approximate decimals, and, for that matter, whole numbers. Let's see how it's done.

Example 8: Round 2.9754

A. 2.9754: Round to the nearest thousandth

Solution: The thousandth place is the 5; look at the next place, 4. If the digit is less than 5 (it is a 4), drop the digit. Answer is 2.975.

B. Round 2.9754 to the nearest hundredth

Solution: 7 is the hundredth place, and the next digit is 5 (which is greater than or equal to 5). So we increase the 7 to an 8. Answer is 2.98.

C. Round 2.9754 to the nearest tenth

Solution: 9 is the tenth place and the next digit is 7 which is greater than 5. Add 1 to the 9 getting 10. Answer is 3.0. Note: The zero remains for accuracy.

D. Round 2.9754 to the nearest whole number

Solution: 2 is the whole part and 9 is the next digit, greater than 5. Round the 2 to a 3. Answer is 3.

Example 9: Round 78649

A. Round 78649 to the nearest ten

Solution: 4 is the 10's place; since 9 is 5 or more, add 1 to the 4 and change the 9 to a 0. Answer is 78,650.

B. Round 78649 to the nearest hundred

Solution: 100's place is 6; next place, 4, is less than 5; replace by a 0 and leave the 6 alone; the remaining place, in this case 9, is replaced by a 0. Answer is 78,600.

C. Round 78649 to the nearest thousand

Solution: 8 is the thousand place; 6 is more than 5; the 8 becomes a 9; replace everything to the right of 9 by a 0. Answer is 79,000.

D. Round 78649 to the nearest ten thousand

Solution: 7 is the ten thousand place; 8 is more than 5; the 7 becomes an 8; all the remaining become 0. Answer is 80,000.

Now, let's go over fractions.

FRACTIONS

The top of a fraction is called the **numerator**; the bottom is the **denominator**.

Rule 1: If the bottoms of two fractions are the same, the bigger the top, the bigger the fraction.

Example 10: Suppose I am a smart first grader. Can you explain to me which is bigger, $\frac{3}{5}$ or $\frac{4}{5}$?

Solution: Suppose we have a pizza pie. Then $\frac{3}{5}$ means we divide a pie into 5 equal parts, and I get 3. And $\frac{4}{5}$ means I get 4 pieces out of 5. So $\frac{3}{5} < \frac{4}{5}$.

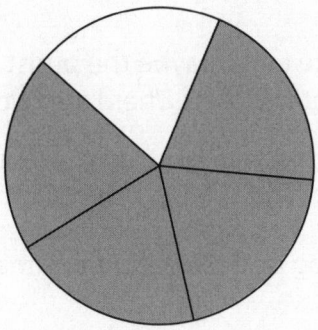

$\frac{4}{5}$

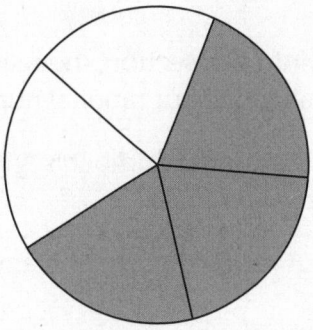

$\frac{3}{5}$

Rule 2: If the tops of two fractions are the same, the bigger the bottom, the smaller the fraction.

Example 11: Which fraction, $\frac{3}{5}$ or $\frac{3}{4}$, is bigger?

Solution: Use another pizza pie example. In comparing $\frac{3}{5}$ and $\frac{3}{4}$, we get the same number of pieces (3). However, if the pie is divided into 4, the pieces are bigger, so $\frac{3}{5} < \frac{3}{4}$.

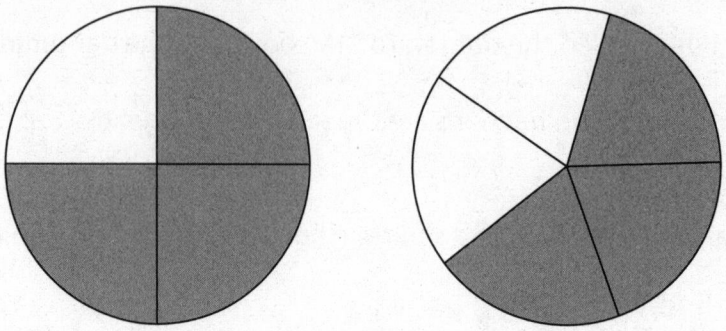

$\frac{3}{4}$ $\frac{3}{5}$

Rule 3: If the tops and bottoms are different, find the least common denominator (LCD) and compare the tops.

Before we get into this section, the teacher in me (and maybe the purist in you) must tell you we really are talking about rational numbers, not fractions. There are two definitions.

Definition 1: A **rational** number is any integer divided by an integer, with the denominator not equaling zero.

Definition 2: A **rational** number is any repeating or terminating decimal.

Note *Technically,* $\frac{\pi}{6}$ *is a fraction but not a rational number. We will use the term* fraction *here instead of* rational number. *If it is negative, we will say "negative fraction."*

Note the following facts about fractions:

- $3 < 4$, but $-3 > -4$. Similarly, $\frac{3}{5} < \frac{4}{5}$, but $-\frac{3}{5} > -\frac{4}{5}$. We will do more of this later.
- A fraction (positive) is bigger than 1 if the numerator is bigger than the denominator.
- A fraction is less than $\frac{1}{2}$ if the denominator is more than twice the numerator.
- To double a fraction, either multiply the numerator by two or divide the denominator by 2.
- Adding the same positive number to the numerator and denominator makes the fraction closer to one.
- Two fractions are **equivalent** if they can be reduced to the same fraction. $\frac{6}{9}$ and $\frac{10}{15}$ are equivalent because they both reduce to $\frac{2}{3}$.

Let's do a few more multiple-choice exercises.

Exercise 3: Which fraction is largest?

A. $\frac{7}{4}$

B. $\frac{11}{8}$

C. $\frac{3}{2}$

D. $\frac{12}{5}$

E. $\frac{3}{100}$

Exercise 4: Which fraction is smallest?

A. $\frac{3}{5}$

B. $\frac{5}{11}$

C. $\frac{7}{13}$

D. $\frac{9}{16}$

E. $\frac{100}{199}$

Exercise 5: Which fraction is largest?

A. $\frac{1}{3}$

B. $\frac{1}{.3}$

C. $\frac{1}{(.3)^2}$

D. $\frac{3}{.1}$

E. $\frac{3}{(.1)^2}$

Exercise 6: If $100 \leq x \leq 10{,}000$ and $.0001 \leq y \leq .01$, the smallest possible value of $\frac{x}{y}$ is

A. 10,000

B. 100,000

C. 1,000,000

D. 10,000,000

E. 1,000,000,000

Exercise 7: Suppose we have $\frac{x+y}{x-y}$, where $8 \leq x \leq 10$ and $2 \leq y \leq 4$. The maximum possible value for this fraction is

A. $\frac{3}{2}$

B. $\frac{5}{3}$

C. $\frac{7}{3}$

D. 3

E. 6

Let's look at the answers.

Answer 3: **D:** It is the only fraction for which the top is more than double the bottom, so it is the only fraction with a value greater than 2.

Answer 4: **B:** It is the only fraction less than one-half; the bottom is more than double the top.

Answer 5: **E:** Answer choice A is less than 1. Comparing the other answer choices, C is bigger than B, and E is bigger than D. Because E has a bigger top and smaller bottom than C, it is larger. We will explore this in greater detail in the next chapter.

Answer 6: **A:** To make a fraction as small as possible, we need a fraction with the smallest top ($x = 100$) and the biggest bottom ($y = .01$), or $\frac{100}{.01} = 10{,}000$.

Answer 7: **D:** This one is not so simple. In this case, the extreme values, minimum (min) and maximum (max), occur at the "ends." We must try $x = 8$ and 10, $y = 2$ and 4, and all combinations of these numbers. In this case, the max occurs when $x = 8$ and $y = 4$.

Adding and Subtracting Fractions

If the denominators are the same, add or subtract the tops, keep the denominator the same, and reduce if necessary.

$$\frac{7}{43} + \frac{11}{43} - \frac{2}{43} = \frac{16}{43}$$

$$\frac{2}{9} + \frac{4}{9} = \frac{6}{9} = \frac{2}{3}$$

$$\frac{a}{m} + \frac{b}{m} - \frac{c}{m} = \frac{a+b-c}{m}$$

There is much more to talk about if the denominators are unlike.

The quickest way to add (or subtract) fractions with different denominators, especially if they contain letters or the denominators are small (but different), is to multiply the top and bottom of each fraction by the least common multiple, or LCM. This consists of three words: *multiple*, *common*, and *least*.

Example 12: What is the LCM of 6 and 8?

Solution: **Multiples** of 6 are 6, 12, 18, 24, 30, 36, 42, 48, 54, 60, 66, 72, 78, . . .

Multiples of 8 are 8, 16, 24, 32, 40, 48, 56, 64, 72, 80, . . .

Common multiples of 6 and 8 are 24, 48, 72, 96, 120, . . .

The **least common multiple** of 6 and 8 is 24.

When adding or subtracting fractions with different denominators, multiply the top and bottom of each fraction by the LCM divided by the denominator of that fraction:

$$\frac{a}{b} - \frac{x}{y} = \left(\frac{a}{b} \times \frac{y}{y}\right) - \left(\frac{x}{y} \times \frac{b}{b}\right) = \frac{ay}{by} - \frac{bx}{by} = \frac{ay - bx}{by}$$

$$\frac{7}{20} - \frac{3}{11} = \left(\frac{7}{20} \times \frac{11}{11}\right) - \left(\frac{3}{11} \times \frac{20}{20}\right) = \frac{7(11) - 3(20)}{20(11)} = \frac{17}{220}$$

On the ACCUPLACER, you must be able to perform these calculations quickly.

Example 13: Find the sums:

Problem	Solution
a. $\frac{3}{8} + \frac{5}{6} =$	$\frac{9}{24} + \frac{20}{24} = \frac{29}{24}$
b. $\frac{3}{4} + \frac{5}{6} =$	$\frac{3}{4} + \frac{5}{6} = \frac{9}{12} + \frac{10}{12} = \frac{19}{12}$
c. $4\frac{3}{8} + 5\frac{5}{6} =$	$4\frac{9}{24} + 5\frac{20}{24} = 9\frac{29}{24} = 9 + 1\frac{5}{24} = 10\frac{5}{24}$

Doing this problem on the ACCUPLACER can take a long time by calculator. It is to your advantage to be able to do problems like this with paper and pencil.

Multiplication of Fractions

To multiply fractions, multiply the numerators and multiply the denominators, reducing as you go. With multiplication, it is *not* necessary to have the same denominators.

$$\frac{3}{7} \times \frac{4}{11} = \frac{12}{77}$$

$$\frac{a}{b} \times \frac{c}{d} = \frac{a \times c}{b \times d}$$

Division of Fractions

To divide fractions, invert the second fraction and multiply, reducing if necessary. To **invert** a fraction means to turn it upside down. The new fraction is called the **reciprocal** of the original fraction. So the reciprocal of $\frac{2}{3}$ is $\frac{3}{2}$; the reciprocal of -5 is $-\frac{1}{5}$; and the reciprocal of a is $\frac{1}{a}$ if $a \neq 0$.

Example 14: Do the following calculations:

Problem	Solution
a. $\frac{3}{4} \div \frac{11}{5} =$	$\frac{3}{4} \times \frac{5}{11} = \frac{15}{44}$
b. $\frac{m}{n} \div \frac{p}{q} =$	$\frac{m}{n} \times \frac{q}{p} = \frac{m \times q}{n \times p}$
c. $\frac{1}{4} \div 5 =$	$\frac{1}{4} \times \frac{1}{5} = \frac{1}{20}$
d. $3\frac{1}{5} \times 1\frac{1}{6} =$	$\frac{16}{5} \times \frac{7}{6} = \frac{8}{5} \times \frac{7}{3} = \frac{56}{15}$ or $3\frac{11}{15}$

Let's do a few problems.

Example 15:

Problem	Solution
a. $\frac{7}{9} - \frac{3}{22} =$	$\frac{127}{198}$
b. $\frac{3}{4} + \frac{5}{6} - \frac{1}{8} =$	$\frac{35}{24}$, or $1\frac{11}{24}$
c. $\frac{3}{10} + \frac{2}{15} - \frac{4}{5} =$	$\frac{-11}{30}$
d. $\frac{1}{4} + \frac{1}{8} + \frac{7}{16} =$	$\frac{13}{16}$
e. $2 + \frac{2}{3} + \frac{2}{9} + \frac{2}{27} =$	$2\frac{26}{27}$, or $\frac{80}{27}$
f. $\frac{5}{24} - \frac{7}{18} =$	$-\frac{13}{72}$

g. $\frac{10}{99} - \frac{9}{100} =$ $\frac{109}{9900}$

h. $\frac{7}{9} \times \frac{5}{3} =$ $\frac{35}{27}$, or $1\frac{8}{27}$

i. $\frac{11}{12} \div \frac{9}{11} =$ $\frac{121}{108}$, or $1\frac{13}{108}$

j. $\frac{5}{9} \times \frac{6}{7} =$ $\frac{10}{21}$

k. $\frac{12}{13} \div \frac{8}{39} =$ $\frac{9}{2}$, or $4\frac{1}{2}$

l. $\frac{10}{12} \div \frac{15}{40} =$ $\frac{20}{9}$, or $2\frac{2}{9}$

m. $\frac{2}{3} \div 12 =$ $\frac{1}{18}$

n. $\frac{2}{3} \times \frac{3}{4} \times \frac{4}{5} \times \frac{5}{6} \times \frac{6}{7} =$ $\frac{2}{7}$

o. $\frac{5}{8} \times \frac{7}{6} \div \frac{35}{24} =$ $\frac{1}{2}$

p. $3\frac{2}{3} + 4\frac{3}{4} =$ $8\frac{5}{12}$

q. $5\frac{1}{6} - 3\frac{5}{6} =$ $1\frac{1}{3}$

r. $7\frac{3}{7} - 2\frac{1}{2} =$ $4\frac{13}{14}$

s. $3\frac{4}{5} \times \frac{3}{38} =$ $\frac{3}{10}$

t. $4\frac{1}{5} \div 8\frac{2}{5} =$ $\frac{1}{2}$

Q Let's do a couple more multiple-choice exercises.

Exercise 8: What number, when multiplied by $\frac{3}{4}$, gives $\frac{7}{8}$?

A. $\frac{21}{32}$
B. $\frac{6}{5}$
C. $\frac{7}{6}$
D. $\frac{8}{7}$
E. $\frac{4}{3}$

Exercise 9: The average (mean) of $\frac{1}{4}$ and $\frac{1}{8}$ is

A. $\frac{1}{12}$
B. $\frac{1}{6}$
C. $\frac{3}{16}$
D. $\frac{5}{32}$
E. $\frac{7}{24}$

A Let's look at the answers.

Answer 8: **C:** We need to solve $\frac{3}{4}x = \frac{7}{8}$. Then $x = \frac{7}{8} \times \frac{4}{3} = \frac{7}{6}$. You may need to be able to do this is your head.

Answer 9: **C:** $\frac{\left(\frac{1}{4} + \frac{1}{8}\right)}{2} = \frac{\left(\frac{3}{8}\right)}{2} = \frac{3}{16}$.

Changing from Decimals to Fractions and Back

To change from a decimal to a fraction, we read it and write it.

Example 16: Change 4.37 to a fraction.

Solution: We read it as 4 and 37 hundredths: $4\frac{37}{100} = \frac{437}{100}$, if necessary. That's it.

Example 17: Change to decimals:

a. $\frac{7}{4}$
b. $\frac{1}{6}$

Solution: For such fractions, the decimal will either terminate or repeat.

a. Divide 4 into 7.0000: $7.0000 \div 4 = 1.75$

b. Divide 6 into 1.0000: $1.0000 \div 6 = .1666\ldots = .1\overline{6}$

The bar over the 6 means it repeats forever; for example, $.3454545\ldots = .3\overline{45}$. The bar over the 45 means 45 repeats forever, but not the 3.

PERCENTAGES

% means hundredths: $1\% = \frac{1}{100} = .01$.

Follow these rules to change between percentages and decimals and fractions:

Rule 1: To change a percentage to a decimal, move the decimal point two places to the left and drop the % sign.

Rule 2: To change a decimal to a percentage, move the decimal point two places to the right and add a % sign.

Rule 3: To change from a percentage to a fraction, divide by 100% and simplify, or change the % sign to $\frac{1}{100}$ and multiply.

Rule 4: To change a fraction to a percentage, first change to a decimal, and then to a percentage.

Example 18: Change 12%, 4%, and .7% to decimals.

Solutions: $12\% = 12.\% = .12$; $4\% = 4.\% = .04$; $.7\% = .007$.

Example 19: Change .734, .2, and 34 to percentages.

Solutions: $.734 = 73.4\%$; $.2 = 20\%$; $34 = 34. = 3400\%$.

Example 20: Change 42% to a fraction.

Solution: $42\% = .42 = \frac{42}{100} = \frac{21}{50}$, or $42\% = 42 \times \frac{1}{100} = \frac{42}{100} = \frac{21}{50}$

Example 21: Change $\frac{7}{4}$ to a percentage.

Solution: $\frac{7}{4} = 1.75 = 175\%$

It is important to know these for the test.

NANUET PUBLIC LIBRARY
149 CHURCH STREET
NANUET, NEW YORK. 10954

Fraction	Decimal	Percentage	Fraction	Decimal	Percentage
$\frac{1}{2}$	.5	50%	$\frac{1}{8}$	.125	$12\frac{1}{2}\%$
$\frac{1}{4}$	.25	25%	$\frac{3}{8}$	.375	$37\frac{1}{2}\%$
$\frac{3}{4}$	.75	75%	$\frac{5}{8}$	.625	$62\frac{1}{2}\%$
$\frac{1}{5}$	.2	20%	$\frac{7}{8}$	.875	$87\frac{1}{2}\%$
$\frac{2}{5}$	.4	40%	$\frac{1}{16}$	.0625	$6\frac{1}{4}\%$
$\frac{3}{5}$	.6	60%	$\frac{3}{16}$	.1875	$18\frac{3}{4}\%$
$\frac{4}{5}$	.8	80%	$\frac{5}{16}$	.3125	$31\frac{1}{4}\%$
$\frac{1}{6}$	$.1\overline{6}$	$16\frac{2}{3}\%$	$\frac{7}{16}$	.4375	$43\frac{3}{4}\%$
$\frac{1}{3}$	$.\overline{3}$	$33\frac{1}{3}\%$	$\frac{9}{16}$	.5625	$56\frac{1}{4}\%$
$\frac{2}{3}$	$.\overline{6}$	$66\frac{2}{3}\%$	$\frac{11}{16}$	.6875	$68\frac{3}{4}\%$
$\frac{5}{6}$	$.8\overline{3}$	$83\frac{1}{3}\%$	$\frac{13}{16}$	.8125	$81\frac{1}{4}\%$
			$\frac{15}{16}$	.9375	$93\frac{3}{4}\%$

If you can't memorize all of these with denominator 16, memorize at least $\frac{1}{16}$ and $\frac{15}{16}$.

If you are good at doing percentage problems, skip this next section. Otherwise, here's a really easy way to do percentage problems. Make the following pyramid:

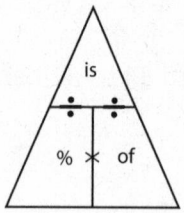

Example 22: What is 12% of 1.3?

Solution:

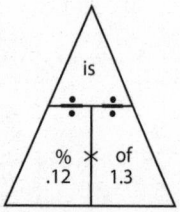

Put .12 in the % box (always change to a decimal in this box) and 1.3 in the "of" box. It tells us to multiply $.12 \times 1.3 = .156$. That's all there is to it.

Example 23: 8% of what is 32?

Solution:

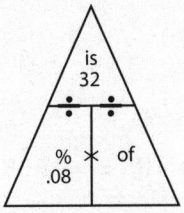

.08 goes in the % box. 32 goes in the "is" box. 32 ÷ .08 = 400.

Example 24: 9 is what % of 8?

Solution:

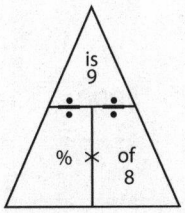

9 goes in the "is" box. 8 goes in the "of" box. 9 ÷ 8 × 100% = 112.5%.

The goal is to be able to do percentage problems without writing the pyramid.

$$\text{Percentage increase} = \frac{\text{New amount} - \text{Original amount}}{\text{Original amount}} \times 100\%$$

$$\text{Percentage decrease} = \frac{\text{Original amount} - \text{New amount}}{\text{Original amount}} \times 100\%$$

Example 25: In ten years, the population increases from 20,000 to 23,000. Find the actual increase and the percentage increase.

Solution: The actual increase is 23,000 − 20,000 = 3,000.

$$\text{Percentage increase} = \frac{23{,}000 - 3{,}000}{23{,}000} = \frac{3{,}000}{23{,}000} \times 100\% = 15\% \text{ increase}$$

Example 26: The cost of producing widgets decreased from 60 cents to 50 cents. Find the actual decrease and percentage decrease.

Solution: $60 - 50 = 10$ cent actual decrease; $\frac{10}{60} = 16\frac{2}{3}\%$ decrease.

$$\text{Percentage decrease} = \frac{60-50}{60} = \frac{10}{60} \times 100\% = 16\tfrac{2}{3}\% \text{ decrease}$$

Note *Percentage increases and decreases are figured on the original amount.*

Example 27: The cost of a \$2000 large-screen TV set is decreased by 30%. If there is a 7% sales tax, how much does it cost?

Solution: $\$2000 \times .30 = \$600$ discount. $\$2000 - \$600 = \$1400$ cost. $\$1400 \times .07$ is \$98. The total price is $\$1400 + \$98 = \$1498$.

Note *If we took 70% (or 100% − 30%) of \$2000, we would immediately get the cost.*

There is an interesting story about why women wear miniskirts in London, England. It seems that the sales tax is $12\frac{1}{2}\%$ on clothes! But children's clothes are tax exempt. A girl's dress is any dress where the skirt is less than 24 inches; so that is why women in London wear miniskirts!

Q **Let's do some multiple-choice exercises.**

Exercise 10: The product of 2 and $\frac{1}{89}$ is:

A. $2\frac{1}{89}$

B. $1\frac{88}{89}$

C. 172

D. $\frac{2}{89}$

E. $\frac{1}{172}$

Exercise 11: $\frac{1}{50}$ of 2% of .02 is

A. .08

B. .008

C. .0008

D. .000008

E. .00000008

Exercise 12: 30% of 20% of a number is the same as 40% of what percentage of the same number?

A. 10

B. $12\frac{1}{2}$

C. 15

D. 18

E. Can't be determined without the number

Exercise 13: A fraction between $\frac{3}{43}$ and $\frac{4}{43}$ is

A. $\frac{1}{9}$

B. $\frac{3}{28}$

C. $\frac{5}{47}$

D. $\frac{7}{86}$

E. $\frac{9}{1849}$

Exercise 14: The reciprocal of $2 - \frac{3}{4}$ is

A. $\frac{1}{2} - \frac{4}{3}$

B. $-\frac{5}{4}$

C. $-\frac{4}{5}$

D. $\frac{5}{4}$

E. $\frac{4}{5}$

Exercise 15: A price increase of 20% followed by a decrease of 20% means the price is

A. Up 4%

B. Up 2%

C. The original price

D. Down 2%

E. Down 4%

Exercise 16: A price decreases 20% followed by a 20% increase. The final price is

A. Up 4%

B. Up 2%

C. The original price

D. Down 2%

E. Down 4%

Exercise 17: A 50% discount followed by another 50% discount gives a discount of

A. 25%
B. 50%
C. 75%
D. 90%
E. 100%

Exercise 18: $83\frac{1}{3}\%$ of $37\frac{1}{2}\%$ is

A. $\frac{1}{2}$
B. $\frac{2}{5}$
C. $\frac{1}{4}$
D. $\frac{5}{16}$
E. $\frac{3}{32}$

Exercise 19: Which is equivalent to .0625?

A. $\frac{3}{8}$
B. $\frac{1}{8}$
C. $\frac{1}{16}$
D. $\frac{1}{18}$
E. $\frac{1}{80}$

Exercise 20: If the fractions $\frac{19}{24}, \frac{1}{2}, \frac{3}{8}, \frac{3}{4}$, and $\frac{9}{16}$ were ordered least to greatest, the middle number would be

A. $\frac{19}{24}$
B. $\frac{1}{2}$
C. $\frac{3}{8}$
D. $\frac{3}{4}$
E. $\frac{9}{16}$

Exercise 21: $\frac{31}{125} =$

A. 0.320
B. 0.310
C. 0.288
D. 0.252
E. 0.248

Let's look at the answers.

Answer 10: **D:** $\frac{2}{1} \times \frac{1}{89} = \frac{2}{89}$.

Answer 11: **D:** $.02 \times .02 \times .02 = .000008$.

Answer 12: **C:** We can use the numbers and forget the percentages.
$(30)(20) = (40)(?)$. $? = 15$.

Answer 13: **D:** $\frac{3}{43} = \frac{6}{86}$; $\frac{4}{43} = \frac{8}{86}$; between is $\frac{7}{86}$. As another example like this one, to get nine numbers between $\frac{3}{43}$ and $\frac{4}{43}$, multiply both fractions, top and bottom, by 10 (one more than nine), and the fractions in between would be $\frac{31}{430}, \frac{32}{430}, \frac{33}{430}$, etc.

Answer 14: **E:** $2 - \frac{3}{4} = \frac{5}{4}$. Its reciprocal is $\frac{4}{5}$.

Answer 15: **E:** Let \$100 be the original price. A 20% increase means a \$20 increase. Then a 20% decrease is $(\$120) \times (.20) = \$24$.
$\$120 - \$24 = \$96$, or a \$4 decrease from the original price. $\frac{\$4}{\$100}$ is a 4% decrease.

Answer 16: **E:** Suppose we have \$100. Increased by 20%, we have \$120. But another 20% less (now on a larger amount) is \$24 less. We are at \$96, down 4% from the original price. Suppose we take 20% off first. We are at \$80. Now 20% up (on a smaller amount) is \$16. We are again at \$96 or again down 4% from the original price.

Answer 17: **C:** \$100 discounted 50% is \$50. 50% of \$50 is \$25, for a 75% discount. This is why after December, a 50% discount followed by 20% is 60%, but sounds like 70% off.

Answer 18: **D:** If you memorized the fraction table, you would recognize this problem as the same as $\left(\frac{5}{6}\right)\left(\frac{3}{8}\right) = \frac{5}{16}$.

Answer 19: **C:** It is easy if you've memorized the fraction table.

Answer 20: **E:** If we changed each number to a fraction with a denominator of 48 (the LCM), we would get $\frac{19}{24} = \frac{38}{48}, \frac{1}{2} = \frac{24}{48}, \frac{3}{8} = \frac{18}{48}, \frac{3}{4} = \frac{36}{48}$, and $\frac{9}{16} = \frac{27}{48}$, and it is easy to see that $\frac{27}{48} = \frac{9}{16}$ is the middle number.

Answer 21: **E:** There are two ways to do this problem. If you recognize that $\frac{31}{125} < \frac{1}{4}$ (= .25) because the bottom is more than 4 times the top, you see that answer choice E is the only one less than $\frac{1}{4}$. Another way to answer this question is to multiply the top and bottom of $\frac{31}{125}$ by 8, and get $\frac{248}{1000} = .248$, which is answer choice E.

Chapter 2 Test

1. $51 + 5.1 + .51 =$
2. $23.4 - 3.49 =$
3. $(.12)(.012) =$
4. $\frac{.12}{.012} =$
5. Change $\frac{9}{8}$ to a decimal and a percent.
6. Change .5% to a decimal and a fraction.
7. Change 3.7 to a fraction and a percent.
8. 14% of 1.4 is what?
9. 12 is what percent of 9?
10. 14% of what is 6.3?
11. Reduce $\frac{26}{65}$.
12. Write $\frac{100}{46}$ as a reduced mixed number.
13. $2\frac{5}{8} + 4\frac{5}{6} =$

14. $12 - 2\frac{3}{7} =$

15. $3\frac{1}{2} \times 4\frac{1}{2} =$

16. $5\frac{1}{4} \div 2\frac{5}{8} =$

17. Find the maximum value of $\frac{x}{y}$ if x and y must be from the set {4, 5, 6, 7}.

18. An $80 radio is discounted 20%. If the state sales tax is 5%, how much does it cost?

19. A stock decreases from $20 a share to $16 a share. Find the percentage decrease.

20. A shirt sold for $45, 75% of the original cost. Find the original cost.

21. Round 1.18 and 1.22 to the nearest tenth; find the product, and round the answer to the nearest tenth.

22. Round $\frac{4}{9}$, 27%, and $\frac{1}{3}$ to the nearest tenth; find the sum and do not round off.

23. Round 4.74449 to the nearest thousandth.

24. Round 7652 to the nearest hundred.

Answers to Chapter 2 Test

1. 51 = 51.0; the answer is 56.61.
2. 19.91.
3. 0.00144.
4. 10.
5. 1.125; 112.5%.
6. 0.005; $\frac{5}{1000} = \frac{1}{200}$.
7. $\frac{37}{10}$, or $3\frac{7}{10}$; 370%.
8. $0.14 \times 1.4 = 0.196$.
9. $\frac{12}{9} \times 100\% = 133\frac{1}{3}\%$.
10. $\frac{6.3}{.14} = 45$.

11. $\frac{2}{5}$.

12. $\frac{50}{23} = 2\frac{4}{23}$.

13. $2\frac{15}{24} + 4\frac{20}{24} = 6\frac{35}{24} = 6 + 1\frac{11}{24} = 7\frac{11}{24}$.

14. $11\frac{7}{7} - 2\frac{3}{7} = 9\frac{4}{7}$.

15. $\frac{7}{2} \times \frac{9}{2} = \frac{63}{4} = 15\frac{3}{4}$.

16. $\frac{21}{4} \times \frac{8}{21} = 2$.

17. $\frac{7}{4}$; select the largest top number and the smallest bottom number.

18. $100\% - 20\% = 80\%$; $.80 \times \$80 = \$64$; $\$64 + .05 \times \$64 = \$67.20$.

19. $20 - 16 = 4$; $\frac{4}{20} \times 100\% = 20\%$.

20. $\frac{45}{.75}$, or $\frac{45}{\left(\frac{3}{4}\right)} = 45 \times \frac{4}{3} = 15 \times 4 = \$60$.

21. $(1.2) \times (1.2) = 1.44$; rounded, the answer is 1.4.

22. $.4 + .3 + .3 = 1.0$; you must leave the zero.

23. 4.744

24. 7,700

Enough! Let's do some algebra now. We will see these topics throughout the book.

Answer to "Bob Asks": 11 fence posts; you need one in the beginning.

CHAPTER 3: *The Power of Exponents*

"A bell rings at even intervals forever. If it rings 6 times in 5 seconds, how many times will it ring in 30 seconds? (Hint: This is like the fence post question at the beginning of Chapter 2.)"

Exponents are a very popular topic on the ACCUPLACER. They are a good test of knowledge and thinking, are short to write, and it is relatively easy to make up new problems. Let's review some basic rules of exponents.

Rule	**Examples**
1. $x^m x^n = x^{m+n}$	$x^6x^4x = x^{11}$ and $(x^6y^7)(x^4y^{10}) = x^{10}y^{17}$
2. $\dfrac{x^m}{x^n} = x^{m-n}$ or $\dfrac{1}{x^{n-m}}$	$\dfrac{x^8}{x^6} = x^2$, $\dfrac{x^3}{x^7} = \dfrac{1}{x^4}$, and $\dfrac{x^4y^5z^9}{x^9y^2z^9} = \dfrac{y^3}{x^5}$
3. $(x^m)^n = x^{mn}$	$(x^5)^7 = x^{35}$
4. $(xy)^n = x^ny^n$	$(xy)^3 = x^3y^3$ and $(x^7y^3)^{10} = x^{70}y^{30}$
5. $\left(\dfrac{x}{y}\right)^n = \dfrac{x^n}{y^n}$	$\left(\dfrac{x}{y}\right)^6 = \dfrac{x^6}{y^6}$ and $\left(\dfrac{y^4}{z^5}\right)^3 = \dfrac{y^{12}}{z^{15}}$
6. $x^{-n} = \dfrac{1}{x^n}$ and $\dfrac{1}{x^{-m}} = x^m$	$2^{-3} = \dfrac{1}{2^3} = \dfrac{1}{8}$, $\dfrac{1}{4^{-3}} = 4^3 = 64$, $\dfrac{x^{-4}y^{-5}z^6}{x^{-6}y^4z^{-1}} = \dfrac{x^6z^6z^1}{x^4y^5y^4} = \dfrac{x^2z^7}{y^9}$, and $\left(\dfrac{x^3}{y^{-4}}\right)^{-2} = \left(\dfrac{y^{-4}}{x^3}\right)^2 = \dfrac{y^{-8}}{x^6} = \dfrac{1}{x^6y^8}$
7. $x^0 = 1, x \neq 0$; 0^0 is indeterminate	$(7ab)^0 = 1$ and $7x^0 = 7(1) = 7$

8. $x^{\frac{p}{r}} = x^{\frac{\text{power}}{\text{root}}}$

Even though either order of computing the power and root gives the same answer, we usually do the root first. $25^{\frac{3}{2}} = \left(\sqrt{25}\right)^3 = 5^3 = 125$;

$$4^{-\frac{3}{2}} = \frac{1}{4^{\frac{3}{2}}} = \frac{1}{\left(\sqrt{4}\right)^3} = \left(\frac{1}{2}\right)^3 = \frac{1}{8}$$

It will save time if you know the following:

$2^2 = 4$	$2^3 = 8$	$2^4 = 16$	$2^5 = 32$	$2^6 = 64$	$2^7 = 128$
$2^8 = 256$	$2^9 = 512$	$2^{10} = 1024$	$3^2 = 9$	$3^3 = 27$	$3^4 = 81$
$3^5 = 243$	$3^6 = 729$	$4^2 = 16$	$4^3 = 64$	$4^4 = 256$	$5^2 = 25$
$5^3 = 125$	$6^2 = 36$	$6^3 = 216$	$7^2 = 49$	$8^2 = 64$	$9^2 = 81$
$10^2 = 100$	$11^2 = 121$	$12^2 = 144$	$13^2 = 169$	$14^2 = 196$	$15^2 = 225$
$16^2 = 256$	$17^2 = 289$	$18^2 = 324$	$19^2 = 361$	$20^2 = 400$	$21^2 = 441$
$22^2 = 484$	$23^2 = 529$	$24^2 = 576$	$25^2 = 625$	$26^2 = 676$	$27^2 = 729$
$28^2 = 784$	$29^2 = 841$	$30^2 = 900$	$31^2 = 961$	$32^2 = 1024$	

Over the years, I have asked my students to memorize these powers for a number of reasons. In math, these powers occur often. They occur when we use the Pythagorean theorem (mostly the squares). They are used going backward to find roots (next chapter). Also, they show number patterns of the squares (look at the last digits of each of the squares). All of these topics are on the ACCUPLACER.

Even though the ACCUPLACER does not ask comparison-type questions, it is necessary to be able to compare x and x^2. Similar results hold for x^3, x^4, and so on.

If $x > 1$, then $x^2 > x$, because $4^2 > 4$.

If $x = 1$, then $x^2 = x$, because $1^2 = 1$.

If $0 < x < 1$, then $x > x^2$, because $\frac{1}{2} > \left(\frac{1}{2}\right)^2 = \frac{1}{4}$!!

If $x = 0$, then $x = x^2$, because $0 = 0^2$.

If $x < 0$, then $x^2 > x$, because the square of a negative is a positive.

Also recall that if $0 < x < 1$, then $\frac{1}{x} > 1$, and if $x > 1$, then $0 < \frac{1}{x} < 1$.

Also if $-1 < x < 0$, then $\frac{1}{x} < -1$, and if $x < -1$, then $-1 < \frac{1}{x} < 0$.

Here are some exponential problems to practice, including some with negative exponents.

Example 1: Simplify the following:

Problem	Solution
a. $(-3a^4bc^6)(-5ab^7c^{10})(-100a^{100}b^{200}c^{2000}) =$	$-1500a^{105}b^{208}c^{2016}$
b. $(10ab^4c^7)^3 =$	$1000a^3b^{12}c^{21}$
c. $(4x^6)^2(10x)^3 =$	$16{,}000x^{15}$
d. $((2b^4)^3)^2 =$	$64b^{24}$
e. $(-b^6)^{101} =$	$-b^{606}$
f. $(-ab^8)^{202} =$	$a^{202}b^{1616}$
g. $\frac{24e^9f^7g^5}{72e^9f^{11}g^7} =$	$\frac{1}{3f^4g^2}$
h. $\frac{(x^4)^3}{x^4} =$	x^8
i. $\left(\frac{m^3n^4}{m^7n}\right)^5 =$	$\frac{n^{15}}{m^{20}}$
j. $\left(\frac{(p^4)^3}{(p^6)^5}\right)^{10} =$	$\frac{1}{p^{180}}$
k. $(-10a^{-4}b^5c^{-2})(4a^{-7}b^{-1}) =$	$\frac{-40b^4}{a^{11}c^2}$
l. $(3ab^{-3}c^4)^{-3} =$	$\frac{b^9}{27a^3c^{12}}$

m. $(3x^4)^{-4}\left(\left(\frac{1}{9x^8}\right)^{-1}\right)^2 =$ 1

n. $(2x^{-4})^2(3x^{-3})^{-2} =$ $\frac{4}{9x^2}$

o. $\left(\frac{(2y^3)^{-2}}{(4x^{-5})}\right)^{-2} =$ $\frac{256y^{12}}{x^{10}}$

Now, let's try some multiple-choice exercises.

Exercise 1: $0 < x < 1$: Arrange in order of smallest to largest: x, x^2, x^3.

A. $x < x^2 < x^3$
B. $x < x^3 < x^2$
C. $x^2 < x < x^3$
D. $x^3 < x^2 < x$
E. $x^3 < x < x^2$

Exercise 2: $-1 < x < 0$: Arrange in order of largest to smallest: x^2, x^3, x^4.

A. $x^4 > x^3 > x^2$
B. $x^4 > x^2 > x^3$
C. $x^3 > x^4 > x^2$
D. $x^2 > x^3 > x^4$
E. $x^2 > x^4 > x^3$

Exercise 3: $0 < x < 1$.

I: $x > \frac{1}{x^2}$

II: $\frac{1}{x^2} > \frac{1}{x^4}$

III: $x - 1 > \frac{1}{x-1}$

Which statements are always true?

A. None
B. I only
C. II only
D. III only
E. All

Exercise 4: $(5ab^3)^3 =$

A. $15ab^6$
B. $75ab^6$
C. $125ab^9$
D. $125a^3b^6$
E. $125a^3b^9$

Exercise 5: $\dfrac{(2x^5)^3(3x^{10})^2}{6x^{15}} =$

A. 1

B. x^{20}

C. $2x^{20}$

D. $12x^{20}$

E. $12x^{210}$

Exercise 6: $\left(\dfrac{12x^6}{24x^9}\right)^3 =$

A. $1728x^{27}$

B. $\dfrac{1}{1728x^{27}}$

C. $\dfrac{1}{6x^9}$

D. $\dfrac{1}{8x^9}$

E. $\dfrac{1}{8x^{27}}$

Exercise 7: $\dfrac{(4x^4)^3}{(8x^6)^2} =$

A. $\dfrac{1}{2}$

B. 1

C. $\dfrac{1}{2}x^{28}$

D. x^{28}

E. $\dfrac{1}{2x}$

Exercise 8: $-1 \le x \le 5$. Where is x^2 located?

A. $-1 \le x^2 \le 5$

B. $0 \le x^2 \le 25$

C. $1 \le x^2 \le 5$

D. $1 \le x^2 \le 10$

E. $1 \le x^2 \le 25$

Exercise 9: $2^m + 2^m =$

A. 2^{m+1}

B. 2^{m+2}

C. 2^{m+4}

D. 2^{m^2}

E. 4^m

Exercise 10: $\dfrac{m^{-5}n^6p^{-2}}{m^{-3}n^9p^0} =$

A. $m^2n^3p^2$

B. $\dfrac{1}{m^2n^3}$

C. $\dfrac{m^2}{n^3p^2}$

D. $\dfrac{1}{m^2n^3p^2}$

E. None of these

Exercise 11: If $8^{2n+1} = 2^{n+18}$, $n =$

A. 3

B. 7

C. 10

D. 13

E. 17

Exercise 12: $p = 4^n$; $4p =$

A. 4^{n+1}

B. 4^{n+2}

C. 3^{n+4}

D. 16^p

E. 64^p

Exercise 13: $y^3 = 64$; $y^{-2} =$

A. -4

B. $-\dfrac{1}{8}$

C. $\dfrac{1}{8}$

D. $\dfrac{1}{16}$

E. $-\dfrac{1}{16}$

Exercise 14: Suppose $x^2 = y^2$.

I: $x = y$.

II: $x = -y$.

III: $x^2 = xy$.

Which statements are always true?

A. None

B. I only

C. II only

D. III only

E. All statements are true

Exercise 15: $3^{-2} =$

A. $\frac{1}{3}$

B. $\frac{1}{6}$

C. $\frac{1}{9}$

D. $-\frac{1}{9}$

E. $-\frac{1}{6}$

Exercise 16: $x^{\frac{3}{4}} = 8; x =$

A. $\frac{32}{3}$

B. 16

C. 64

D. 256

E. 1024

Exercise 17: n is an integer, and $(-2)^{6n} = 8^{n+4}; n =$

A. 2

B. 3

C. 4

D. 6

E. 8

Let's look at the answers.

Answer 1: **D:** If $0 < x < 1$, the higher the power, the smaller the number.

Answer 2: **E:** Take, for example, $x = -\frac{1}{2}$. $\left(-\frac{1}{2}\right)^2 = \frac{1}{4}$; $\left(-\frac{1}{2}\right)^3 = -\frac{1}{8}$; $\left(-\frac{1}{2}\right)^4 = \frac{1}{16}$. Be careful! This exercise asks for largest to smallest. Notice that x^3 has to be the smallest because it is the only negative number, so the answer choices are reduced to two, B and E.

Answer 3: **D:** Statement I is false: $0 < x < 1$, so $\frac{1}{x} > 1$ and $\frac{1}{x^2} > 1$ also. Statement II is false: $x^2 > x^4$, so $\frac{1}{x^2} < \frac{1}{x^4}$. Statement III is true: $0 < x < 1$; so $-1 < x - 1 < 0$; this means $\frac{1}{x-1} < -1$.

Answer 4: **E:** $5^3a^3(b^3)^3 = 125a^3b^9$.

Answer 5: D: $\left(\frac{8 \times 9}{6}\right) x^{15+20-15} = 12x^{20}$.

Answer 6: D: $\left(\frac{1}{2x^3}\right)^3 = \frac{1}{8x^9}$.

Answer 7: B: The numerator and denominator of the fraction each equal $64x^{12}$.

Answer 8: B: This is very tricky because 0 is between -1 and 5, and $0^2 = 0$.

Answer 9: A: This is one of the few truly hard problems because it is an addition problem:

$$2^m + 2^m = 1 \times 2^m + 1 \times 2^m = 2 \times 2^m = 2^1 \times 2^m = 2^{m+1}.$$

Similarly, $3^m + 3^m + 3^m = 3^{m+1}$ and four 4^m terms added equal 4^{m+1}.

Answer 10: D: $\frac{m^{-5}n^6p^{-2}}{m^{-3}n^9p^0} = \frac{m^3n^6}{m^5n^9p^2} = \frac{1}{m^2n^3p^2}$.

Answer 11: A: $8^{2n+1} = (2^3)^{2n+1} = 2^{n+18}$. If the bases are equal, the exponents must be equal. $3(2n+1) = n + 18$; so $n = 3$.

Answer 12: A: $p = 4^n$; $4p = 4(4^n) = 4^1 4^n = 4^{n+1}$.

Answer 13: D: $y^3 = 64$; $y = 4$; $y^{-2} = \left(\frac{1}{4}\right)^2 = \frac{1}{16}$.

Answer 14: A: None! Because $x = y$ or $x = -y$, all of the statements are true sometimes, but none is always true.

Answer 15: C: $3^{-2} = \left(\frac{1}{3}\right)^2 = \frac{1}{9}$.

Answer 16: B: $x^{\frac{3}{4}} = 8$; $x = \left(x^{\frac{3}{4}}\right)^{\frac{4}{3}} = 8^{\frac{4}{3}} = \left(\sqrt[3]{8}\right)^4 = 2^4 = 16$.

Answer 17: C: We can ignore the minus sign, because both sides must be positive. $2^{6n} = 8^{n+4} = 2^{3(n+4)}$. So $6n = 3n + 12$, and $n = 4$.

This chapter seems the only good place to introduce the topic of **scientific notation**. On your calculator, if you see 3.176 and to the right of it you see 27, the number is written in scientific notation. It means 3.176×10^{27}.

A number in scientific notation is of the form $n \times 10^m$, where n is at least 1 but less than 10.

Example 2: Write 567000 in scientific notation.

Solution: Put the decimal place after the first digit 5.67; then count the number of places to the right (of 5) to the end of the original number. This is the power of 10, or m, so the answer is 5.67×10^5. It is a positive exponent because we are counting to the right and because the original number is larger than 10.

Example 3: Write 0.000000097 in scientific notation.

Solution: Again, put the decimal after the first digit, 9.7; then count the number of places from the decimal point after the 9 left to where it was in the original number. It is 8 places; the exponent of 10 is negative here because it is to the left and the entire number is less than 1. The answer is $0.000000097 = 9.7 \times 10^{-8}$.

Example 4: Compute the following in scientific notation: $\dfrac{60000 \times .008}{.00002 \times 200000000}$

Solution: Write all the numbers in scientific notation : $\dfrac{6 \times 10^4 \times 8 \times 10^{-3}}{2 \times 10^{-5} \times 2 \times 10^8}$.

Separate the number parts from the powers of ten. As before, bring negative exponents in the top to the bottom and the bottom to the top, making all of them positive. Then do the math.

$$\frac{6 \times 8}{2 \times 2} \times \frac{10^4 \times 10^5}{10^3 \times 10^8} = 12 \times \frac{10^9}{10^{11}} = 12 \times 10^{-2} = 1.2 \times 10^1 \times 10^{-2} = 1.2 \times 10^{-1}$$

Chapter 3 Test

Simplify the expressions in questions 1–5.

1. $(-3x^{20}y^{30})(-10x^4y)$
2. $\dfrac{(4x^3)^2(10x^4)^3}{(2x^3)^5}$
3. $x^{3a+2}x^{4a+7}$
4. $6^8 6^6 6$
5. $(4x)^0 + 4x^0 + 4^0x^0$

Evaluate the expressions in questions 6–10.

6. $\dfrac{4^{-3}}{2^{-5}}$
7. $8^{-\frac{5}{3}}$
8. $\dfrac{1}{100^{-\frac{7}{2}}}$
9. $3^{-3} - \dfrac{1}{4^{-\frac{3}{2}}}$
10. $4^6 4^{-8} 4^{\frac{5}{2}}$

Simplify the expressions in questions 11–14.

11. $\left(\dfrac{(m^4)}{(m^{-3})}\right)^{-5}$
12. $\dfrac{4x^8y^{-4}z^{-5}}{12x^{-2}y^{-1}z^5}$
13. $(4xy^{-3}z^{-5})(-5x^{-5}y^2z^7)$
14. $x^{\frac{3}{2}}x^{\frac{2}{3}}x^{-\frac{1}{6}}$
15. Simplify $3^n + 3^n + 3^n$.
16. If $x^{-2} = \dfrac{1}{16}$, $x^3 = ?$
17. If $4^{3x+2} = 8^{4x+3}$, $x = ?$
18. Simplify $\dfrac{(x^m)^{m+1}}{x^m}$.
19. Simplify $(4xy^{-4}z^5a^{-6}b^{-1})^{-5}$.
20. In scientific notation, $\dfrac{(.0008)(800000000)}{.002} =$

Answers to Chapter 3 Test

1. $30x^{24}y^{31}$.
2. $\dfrac{(16x^6)(1000x^{12})}{32x^{15}} = 500x^3$.
3. x^{7a+9}.
4. 6^{15}.
5. $1 + 4 + 1 = 6$.
6. $\dfrac{2^5}{4^3} = \dfrac{32}{64} = \dfrac{1}{2}$.
7. $8^{-\frac{5}{3}} = \dfrac{1}{8^{\frac{5}{3}}} = \dfrac{1}{\left(\sqrt[3]{8}\right)^5} = \dfrac{1}{2^5} = \dfrac{1}{32}$.
8. $\dfrac{1}{100^{-\frac{7}{2}}} = 100^{\frac{7}{2}} = \left(\sqrt{100}\right)^7 = 10{,}000{,}000$.
9. $3^{-3} - \dfrac{1}{4^{-\frac{3}{2}}} = \dfrac{1}{27} - 8 = -7\dfrac{26}{27}$.
10. $4^{\frac{1}{2}} = 2$.
11. $\dfrac{m^{-20}}{m^{15}} = \dfrac{1}{m^{35}}$.
12. $\dfrac{4x^8x^2y^1}{12y^4z^5z^5} = \dfrac{x^{10}}{3y^3z^{10}}$.
13. $-20\dfrac{z^2}{x^4y}$.
14. x^2 (just add the exponents).
15. $1 \times 3^n + 1 \times 3^n + 1 \times 3^n = 3 \times 3^n = 3^1 3^n = 3^{n+1}$.
16. $\dfrac{1}{x^2} = \dfrac{1}{16}$; $x^2 = 16$; so $x = \pm 4$; and $x^3 = \pm 64$.
17. $4^{3x+2} = 8^{4x+3}$; $(2^2)^{2x+3} = (2^3)^{4x+3}$; so $2(2x + 3) = 3(4x + 3)$; $x = -\dfrac{5}{6}$.
18. $\dfrac{x^{m^2+m}}{x^m} = \dfrac{x^{m^2}x^m}{x^m} = x^{m^2}$.

19. $4^{-5}x^{-5}y^{20}z^{-25}a^{30}b^{5} = \dfrac{a^{30}b^{5}y^{20}}{1{,}024x^{5}z^{25}}$. (It is prettier to write letters alphabetically and so that you don't forget one, but any order is correct!)

20. $\dfrac{8 \times 10^{-4} \times 8 \times 10^{8}}{2 \times 10^{-3}} = \dfrac{8 \times 8}{2} \times 10^{-4}10^{8}10^{3} = 32 \times 10^{7} = 3.2 \times 10^{8}$.

Now let's go to a radical chapter.

Answer to "Bob Asks": 31!!! You start the time from the first ring. So there are two rings in 1 second, 3 rings in 2 seconds, 6 rings in 5 seconds, and 31 rings in 30 seconds.

CHAPTER 4: *The Most Radical Chapter of All*

"If one hamburger and one hot dog cost $4.00, and one hamburger and three hot dogs cost $7.00, how much does one hamburger cost?"

The square root symbol ($\sqrt{\ }$) is probably the one symbol most people actually like, even for people who don't like math. How else can you explain the square root symbol on a business calculator? I have yet to find a use for it. Here are some basic facts about **square roots** that you should know.

1. You should know the following square roots:

   $\sqrt{0} = 0$ $\quad\sqrt{1} = 1$ $\quad\sqrt{4} = 2$ $\quad\sqrt{9} = 3$ $\quad\sqrt{16} = 4$ $\quad\sqrt{25} = 5$

   $\sqrt{36} = 6$ $\quad\sqrt{49} = 7$ $\quad\sqrt{64} = 8$ $\quad\sqrt{81} = 9$ $\quad\sqrt{100} = 10$

   The numbers under the radicals (square root signs) are called perfect squares because their square roots are whole numbers.

2. $\sqrt{2} \approx 1.4$ (actually it is 1.414 . . .), and $\sqrt{3} \approx 1.73$ (actually it is 1.732 . . . , the year George Washington was born).

3. $\sqrt{\dfrac{a}{b}} = \dfrac{\sqrt{a}}{\sqrt{b}}$, so $\quad\sqrt{\dfrac{25}{9}} = \dfrac{5}{3}$ $\quad\sqrt{\dfrac{7}{36}} = \dfrac{\sqrt{7}}{6}$ $\quad\sqrt{\dfrac{45}{20}} = \sqrt{\dfrac{9}{4}} = \dfrac{3}{2}$.

4. A method of simplification involves finding all the prime factors:

   $\sqrt{200} = \sqrt{(2)(2)(2)(5)(5)} = (2)(5)\sqrt{2} = 10\sqrt{2}$. We can also simplify by recognizing perfect squares: $\sqrt{200} = \sqrt{100 \times 2} = 10\sqrt{2}$. My choice is the first. You have to decide which method works best for you.

5. Adding and subtracting radicals involves combining like radicals:

   $4\sqrt{7} + 5\sqrt{11} + 6\sqrt{7} - 9\sqrt{11} = 10\sqrt{7} - 4\sqrt{11}$

   You cannot combine unlike radicals.

6. Multiplication of radicals follows this rule: $a\sqrt{b} \times c\sqrt{d} = ac\sqrt{bd}$.

   Therefore, $3\sqrt{13} \times 10\sqrt{7} = 30\sqrt{91}$, and

   $10\sqrt{8} \times 3\sqrt{10} = 10 \times 3\sqrt{(2 \times 2 \times 2) \times (2 \times 5)} = 3\sqrt{(2 \times 2) \times (2 \times 2) \times 5} =$

   $10 \times 3 \times 2 \times 2 \times \sqrt{5} = 120\sqrt{5}$.

7. If a single radical appears in the denominator of a fraction, rationalize the denominator (change it to a nonradical) by multiplying both numerator and denominator by the radical:

$$\frac{20}{7\sqrt{5}} = \frac{20}{7\sqrt{5}} \times \frac{\sqrt{5}}{\sqrt{5}} = \frac{20\sqrt{5}}{35} = \frac{4\sqrt{5}}{7} \text{ and } \frac{7}{\sqrt{45}} = \frac{7}{3\sqrt{5}} \times \frac{\sqrt{5}}{\sqrt{5}} = \frac{7\sqrt{5}}{15}$$

8. If the denominator has two terms, one or both of the terms having a square root, multiply the numerator and denominator by its **conjugate**, the same term with a different sign between.

$$\text{So } \frac{3}{7+\sqrt{5}} = \frac{3}{7+\sqrt{5}} \times \frac{7-\sqrt{5}}{7-\sqrt{5}} = \frac{21-3\sqrt{5}}{49-5} = \frac{21-3\sqrt{5}}{44}$$

This will make the denominator a whole number because $(a - b)(a + b) = a^2 - b^2$, and the difference of two squares is a whole number. More on this in Chapter 5.

9. If $c, d > 0$, $\sqrt{c} + \sqrt{d} > \sqrt{c+d}$. Why? If we square the right side, we get $c + d$. If we square the left side, we get $c + d +$ a middle term $(2\sqrt{cd})$.

10. You should know the following **cube roots**:

$\sqrt[3]{1} = 1$	$\sqrt[3]{8} = 2$	$\sqrt[3]{27} = 3$	$\sqrt[3]{64} = 4$	$\sqrt[3]{125} = 5$
$\sqrt[3]{-1} = -1$	$\sqrt[3]{-8} = -2$	$\sqrt[3]{-27} = -3$	$\sqrt[3]{-64} = -4$	$\sqrt[3]{-125} = -5$

11. The square root varies according to the value of the radicand:

If $a > 1$, $a > \sqrt{a}$. For example, $9 > \sqrt{9}$.

If $a = 1$, $a = \sqrt{a}$ because the square root of 1 is 1.

If $0 < a < 1$, $a < \sqrt{a}$. When we take the square root of a positive number, it becomes closer to 1 than the number, so $\sqrt{\frac{1}{4}} = \frac{1}{2} > \frac{1}{4}$.

If $a = 0$, $a = \sqrt{a}$ because the square root of 0 is 0.

If $a > 0$ and $\sqrt{a} < 1$, then $\frac{1}{\sqrt{a}} > 1$. Also, if $\sqrt{a} > 1$, then $\frac{1}{a} < 1$.

12. The square root or any even root of a negative number is imaginary (undefined).

13. The cube root (or any odd root) of a positive number is positive, of a negative number is negative, and any root of 0 is 0. All are always defined (not imaginary).

Note *$\sqrt{9} = 3$, $-\sqrt{9} = -3$, but $\sqrt{-9}$ is imaginary. The equation $x^2 = 9$ has two solutions, $\pm\sqrt{9}$, or ± 3, which stands for both $+3$ and -3.*

Let's do some practice problems.

Example 1: Simplify the following radical expressions:

Problem	Solution
a. $\sqrt{45}$	$3\sqrt{5}$
b. $5\sqrt{72}$	$30\sqrt{2}$
c. $4\sqrt{2} + 5\sqrt{3} - \sqrt{2} - 8\sqrt{3}$	$3\sqrt{2} - 3\sqrt{3}$ or $3(\sqrt{2} - \sqrt{3})$
d. $5\sqrt{8} + 6\sqrt{27} + 9\sqrt{12}$	$10\sqrt{2} + 36\sqrt{3}$
e. $(3\sqrt{7})(5\sqrt{11})$	$15\sqrt{77}$
f. $(10\sqrt{21})(4\sqrt{3})$	$120\sqrt{7}$
g. $(4\sqrt{24})(5\sqrt{8})$	$160\sqrt{3}$
h. $\sqrt{\frac{81}{4}}$	$\frac{9}{2}$ or $4\frac{1}{2}$
i. $\sqrt{\frac{11}{49}}$	$\frac{\sqrt{11}}{7}$ lucky answer
j. $\sqrt{\frac{27}{75}}$	$\frac{3}{5}$
k. $\sqrt{\frac{10}{72}}$	$\frac{\sqrt{5}}{6}$
l. $\frac{3}{\sqrt{5}}$	$\frac{3\sqrt{5}}{5}$
m. $\frac{32}{\sqrt{8}}$	$8\sqrt{2}$
n. $\frac{4}{\sqrt{11} - 2}$	$\frac{4(\sqrt{11} + 2)}{7}$
o. $\frac{\sqrt{2}}{\sqrt{11} + \sqrt{5}}$	$\frac{\sqrt{22} - \sqrt{10}}{6}$

Let's try a few multiple-choice exercises.

Exercise 1: $\left(\sqrt{12}+\sqrt{27}\right)^2 =$

A. 15
B. 39
C. 75
D. 225
E. 675

Exercise 2: Suppose $0 < a < 1$.

I: $a^2 > \sqrt{a}$

II: $\sqrt{a} > \sqrt{a^3}$

III: $\sqrt{a} > \dfrac{1}{\sqrt{a^7}}$

A. I is correct
B. II is correct
C. III is correct
D. I and III are correct
E. II and III are correct

Exercise 3: $c = \left(\dfrac{1}{17}\right)^2 - \sqrt{\dfrac{1}{17}}$. Which answer choice is true for c?

A. $c < -2$
B. $-2 < c < -1$
C. $-1 < c < 0$
D. $0 < c < 1$
E. $1 < c < 2$

Exercise 4: $0 < m < 1$. Arrange in order, smallest to largest, $a = \dfrac{1}{m}$, $b = \dfrac{1}{m^2}$, $c = \dfrac{1}{\sqrt{m}}$.

A. $a < b < c$
B. $a < c < b$
C. $b < c < a$
D. $b < a < c$
E. $c < a < b$

Exercise 5: If $\sqrt[3]{-87} = x$:

A. $-10 < x < -9$

B. $-9 < x < -8$

C. $-5 < x < -4$

D. $-4 < x < -3$

E. x is undefined

Exercise 6: Simplified, $\dfrac{\sqrt{m}}{\sqrt{m} - \sqrt{n}} =$

A. $\dfrac{m - \sqrt{n}}{m - n}$

B. $\dfrac{m + \sqrt{n}}{m + n}$

C. $\dfrac{m + \sqrt{mn}}{m - n}$

D. $\dfrac{m - \sqrt{mn}}{m + n}$

E. 1

Let's look at the answers.

Answer 1: **C:** $\sqrt{12} = \sqrt{2 \times 2 \times 3} = 2\sqrt{3}$ and $\sqrt{27} = \sqrt{3 \times 3 \times 3} = 3\sqrt{3}$. Adding, we get $5\sqrt{3}$. Squaring, we get $25\sqrt{9} = 25 \times 3 = 75$.

Answer 2: **B:** Let's look at the statements one by one.

Statement I: If we square a number between 0 and 1, we make it closer to 0. If we take the square root of the same number, we make it closer to 1. Statement I is wrong.

Statement II: From the previous chapter, if $0 < a < 1$, $a > a^3$. So are its square roots. So statement II is true.

Statement III: $\sqrt{a} < 1$. $a^7 < 1$. So $\sqrt{a^7} < 1$. Then $\dfrac{1}{\sqrt{a^7}} > 1$. Statement III is false.

Answer 3: **C:** $\dfrac{1}{17}$ squared is less than $\dfrac{1}{17}$, the square root is more than $\dfrac{1}{17}$, and both numbers are between 0 and 1. When we subtract a larger from a smaller number, the sign is negative, and here the numbers are fractions, so $-1 < c < 0$.

Answer 4: **E:** Often, these comparison problems are figured out by using an example. If we take $m = \frac{1}{4}$, we see that $\sqrt{m} > m > m^2$. That makes $\frac{1}{\sqrt{m}} < \frac{1}{m} < \frac{1}{m^2}$, or $c < a < b$.

Answer 5: **C:** Because $\sqrt[3]{-125} = -5$ and $\sqrt[3]{-64} = -4$, and because -87 is between -125 and -64, the cube root of -87 must be between -5 and -4.

Answer 6: **C:** $\frac{\sqrt{m}}{\sqrt{m} - \sqrt{n}} \times \frac{\sqrt{m} + \sqrt{n}}{\sqrt{m} + \sqrt{n}} = \frac{m + \sqrt{mn}}{m - n}$. Note that answer choice E couldn't be correct because if the fraction was not equal to 1 before simplification, it couldn't be equal to 1 after simplification.

In many colleges, what you need to know about radicals and how much you need to know in taking further math courses is what was just shown. However, some colleges require more. Here we present some examples of additional facts about radicals from the beginning of this chapter. They involve letters and cube roots, fourth roots, and so on. Following these examples will be some practice exercises involving radicals.

Example 2: Simplify: $4\sqrt{27x^9}$.

Solution: Write the expression using prime factors and group them in pairs where possible. $4\sqrt{(3 \times 3) \times 3 \times (x \times x) \times (x \times x) \times (x \times x) \times (x \times x) \times x} = 12x^4\sqrt{3x}$.

Note *As mentioned in fact #4 at the beginning of this chapter, I think this is the easiest way to simplify radicals with terms in exponential form inside the radical.*

Looking at this example, we see that $\sqrt{x^9} = x^4\sqrt{x}$. Think that the square root is the "2th" root. If you divide 2 into 9, you get 4 with remainder of 1, represented by the x^4 term outside the radical and the remaining x inside the radical.

Use fact #4 to simplify the expressions in Examples 3–5.

Example 3: Simplify: $4\sqrt{3x} + 5\sqrt{6y} + 8\sqrt{3x} - 7\sqrt{6y}$.

Solution: $12\sqrt{3x} - 2\sqrt{6y}$.

Example 4: Simplify: $5x\sqrt{3y} - 9x\sqrt{3y}$.

Solution: $-4x\sqrt{3y}$

Example 5: Simplify: $5\sqrt{8x^5} + 7x\sqrt{18x^3} + 10x^2\sqrt{32x}$.

Solution: The terms can be rewritten as $10x^2\sqrt{2x} + 21x^2\sqrt{2x} + 40x^2\sqrt{2x} = 71x^2\sqrt{2x}$.

Example 6: Multiply and simplify: $5a^2b^3\sqrt{6a^4b^5} \times 2a^2\sqrt{8a^7b}$.

Solution: Use fact #6 for the multiplication: $10a^4b^3\sqrt{48a^{11}b^6} = (10a^4b^3)(4a^5b^3\sqrt{3a}) = 40a^9b^6\sqrt{3a}$.

Example 7: Simplify: $\dfrac{4a^3b^2}{\sqrt{12a^5b^9}}$.

Solution: Use fact #7 to eliminate the radical in the denominator (rationalize the fraction): $\dfrac{4a^3b^2}{2a^2b^4\sqrt{3ab}} \times \dfrac{\sqrt{3ab}}{\sqrt{3ab}} = \dfrac{4a^3b^2\sqrt{3ab}}{2a^2b^4(3ab)} = \dfrac{2\sqrt{3ab}}{3b^3}$.

Note *In one of the more puzzling aspects of math, when you see the word "rationalize," you always rationalize the denominator. However, in calculus you will never rationalize the denominator. You will always rationalize the numerator. The reason is that before calculators, it was much more difficult, took longer, and was less accurate to determine the value of* $\dfrac{1}{\sqrt{2}}$ *than it was to determine the value of* $\dfrac{\sqrt{2}}{2}$.
Whenever mathematicians make very slow changes in methodology, it is always good. Unfortunately, after 1985, the changes were the most rapid. (Translation: *There is a lot of bad stuff around now.)*

Example 8: Simplify: $\dfrac{\sqrt{a}+\sqrt{b}}{\sqrt{a}-\sqrt{b}}$.

Solution: Use fact #8. $\dfrac{\sqrt{a}+\sqrt{b}}{\sqrt{a}-\sqrt{b}} \times \dfrac{\sqrt{a}+\sqrt{b}}{\sqrt{a}+\sqrt{b}} = \dfrac{a+2\sqrt{ab}+b}{a-b}$.

Example 9: Simplify: $\sqrt[3]{a^{20}}$.

Solution: $a^6\sqrt[3]{a^2}$.

Note *If we divide 3 into 20, we get 6 with a remainder of 2, which are the exponents of a.*

Example 10: Simplify: $\sqrt[10]{x^{86}}$.

Solution: $x^8\sqrt[10]{x^6}$

Example 11: Simplify: $\sqrt[3]{32x^5}$.

Solution: $x\sqrt[3]{(2 \times 2 \times 2) \times 2 \times 2 \times x^2} = 2x\sqrt[3]{4x^2}$.

Example 12: Perform the following operations:

Problem	Solution
a. Simplify: $\sqrt{16w^3x}$.	$4w\sqrt{wx}$
b. Simplify: $\sqrt{24u^6v^9}$.	$2u^3v^4\sqrt{6v}$
c. Simplify: $5x^2\sqrt{20x^{101}}$.	$10x^{52}\sqrt{5x}$
d. Simplify: $\sqrt[3]{72z^{32}}$.	$2z^{10}\sqrt[3]{9z^2}$
e. Simplify and combine: $5x^3\sqrt{28x^5} + 7x^4\sqrt{63x^3}$.	$10x^5\sqrt{7x} + 21x^5\sqrt{7x} = 31x^5\sqrt{7x}$
f. Simplify: $\sqrt[3]{x^{100}}$.	$x^{33}\sqrt[3]{x}$
g. Simplify: $\sqrt[5]{y^{77}}$.	$y^{15}\sqrt[5]{y^2}$
h. Simplify: $10y\sqrt[3]{108y^{14}}$.	$30y^5\sqrt[3]{4y^2}$
i. Multiply and simplify: $(2s^4t^{10}\sqrt{6s^3t})(10s^2t^2\sqrt{2st^2})$.	$40s^8t^{13}\sqrt{3t}$
j. Rationalize: $\dfrac{4x^7}{\sqrt{72x^{21}}}$.	$\dfrac{\sqrt{2x}}{3x^4}$
k. Rationalize: $\dfrac{4u}{7-\sqrt{u}}$.	$\dfrac{4u(7+\sqrt{u})}{49-u}$
l. Rationalize: $\dfrac{\sqrt{y}-\sqrt{z}}{\sqrt{y}+\sqrt{z}}$.	$\dfrac{y-2\sqrt{yz}+z}{y-z}$

Let's look at imaginary numbers.

IMAGINARY (COMPLEX) NUMBERS

For this particular book, this is the right time to discuss imaginary (or complex) numbers. Even though this is only a few pages and for most people very simple, it is enough about the subject to take you at least past the entire basic calculus sequence in college, if not further.

Once upon a time, we had the counting numbers (natural numbers) 1, 2, 3, 4, ... But mathematicians wanted an answer to the equation $x + 6 = 6$. So 0 was invented. (Actually, 0 was invented in India around the seventh or eighth century. At that time, the Indians were the great mathematicians of the world. Today, India is still pretty darn good!) Then mathematicians wanted even more. They wanted the answer to the equation $x + 5 = 3$. So negative integers were invented. This was still not enough. Mathematicians wanted an answer to the equations $3x = 7$ and $5x = -2$. So rational numbers were invented. Still, they wanted an answer to the equation $x^2 = 41$. So irrational numbers, such as $\pm\sqrt{41}$, were invented. Finally, mathematicians wanted an answer to the equation $x^2 = -1$. Alas! There is no real number such that $x^2 = -1$. So they invented i to stand for the "imaginary" number $\sqrt{-1}$. This was almost complete, but not quite enough. **Complex numbers** were invented. A **complex number** is of the form $a + bi$, where a and b are real numbers, and $i = \sqrt{-1}$.

Two complex numbers are equal if their real parts are equal and their imaginary parts are equal.

$$a + bi = c + di \text{ if } a = c \text{ and } b = d$$

Complex numbers cannot be graphed on a real graph. So instead of x and y axes, there is an x and a yi axis. On the axes below, we have graphed the point $-8 + 6i$.

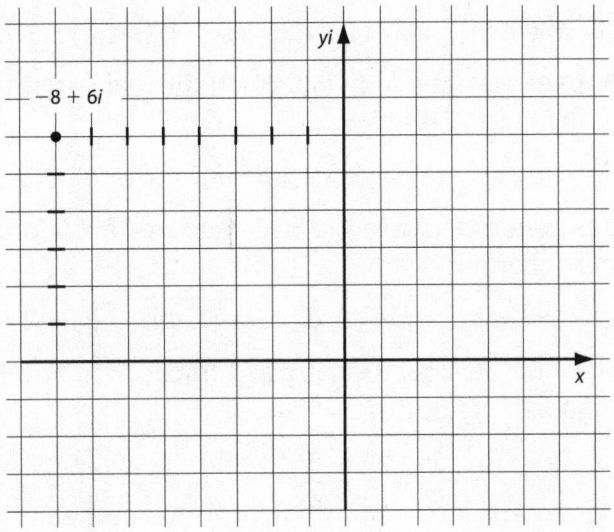

There are only a few things we need to know here about complex numbers.

The distance of the number $a + bi$ from the origin $(0 + 0i)$ is $|a + bi| = \sqrt{a^2 + b^2}$. This is also called its **magnitude**. It is similar to what we get for real numbers by using the distance formula.

Example 13: What is the distance of the point $(6 - 8i)$ from the origin?

Solution: $|6 - 8i| = \sqrt{6^2 + (-8)^2} = 10.$

Note *The distance is always a positive real number and* i *is not involved in the square root.*

We also need to know how to raise i to any power. We see that $i = \sqrt{-1}$, so $i^2 = -1$. $i^3 = i^2 i = -i$; $i^4 = i^2 i^2 = (-1)(-1) = 1$; $i^5 = i^4 i = i$, etc. So i is cyclic, and the cycle is 4. This means the value of i raised to a power repeats itself every fourth power.

Example 14: Simplify i^{1359}.

Solution: To find what part of the i^n cycle this is, we divide 1359 by 4. We are not interested in the answer—only the remainder. The answer is 339 R 3. Because the remainder is 3, the answer is the same as i^3, or $-i$.

To summarize, to find the value of i raised to any power, we divide the exponent by 4. If the remainder is 1, the answer is i. If the remainder is 2, the answer is -1. If the remainder is 3, the answer is $-i$. If the remainder is 0, the answer is 1.

To add, subtract, multiply, and divide complex numbers, follow these rules:

1. To add (subtract) complex numbers, add (subtract) the real parts and add (subtract) the imaginary parts.

$$(a + bi) \pm (c + di) = (a + c) \pm (b + d)i.$$

2. To multiply two complex numbers, use the FOIL method. If you forget what the FOIL method is, see the next chapter.

$$(a + bi)(c + di) = ac + adi + bci + bdi^2 = ac + adi + bci + bd(-1) = ac - bd + (ad + bc)i$$

3. To divide two complex numbers, set them up as a fraction, and multiply the top and bottom by the conjugate of the bottom.

$$\frac{a + bi}{c + di} = \frac{a + bi}{c + di} \times \frac{c - di}{c - di} = \frac{ac + bd + i(bc - ad)}{c^2 + d^2}$$

Note *The* ***conjugate of a complex number*** *is similar to the concept of conjugates with radicals, namely, the same expression with a different sign between the terms.*

Note *The final denominator is always a positive real number. It has a plus sign between the terms because it is* $c^2 - (di)^2 = c^2 - (-d^2) = c^2 + d^2$.

Example 15: Let $Y = 8 + 2i$ and $Z = 6 + 5i$. Find **a.** $Y + Z$; **b.** $Y - Z$; **c.** YZ; **d.** $\frac{Y}{Z}$.

Solution:

a. $Y + Z = 14 + 7i$

b. $Y - Z = 2 - 3i$

c. $YZ = 48 + 40i + 12i + 10i^2 = 38 + 52i$

d. $$\frac{Y}{Z} = \frac{8 + 2i}{6 + 5i} \times \frac{6 - 5i}{6 - 5i} = \frac{48 - 40i + 12i - 10i^2}{36 - 30i + 30i - 25i^2} = \frac{58 - 28i}{61}$$
$$= \frac{58}{61} - \frac{28}{61}i$$

It is essential when taking the square root of a negative number that you write the square root of a negative number in terms of i before proceeding.

Let us see why. When we wrote the rules for radicals, we were assuming positive or nonnegative numbers. In symbols,

$\sqrt{a} \times \sqrt{b} = \sqrt{ab}$ if $a, b \geq 0$, and $\sqrt{\frac{a}{b}} = \frac{\sqrt{a}}{\sqrt{b}}$ if $a \geq 0$ and $b \geq 0$.

Let us see what happens if this is ignored.

$\sqrt{-2} \times \sqrt{-8} = i\sqrt{2} \times i\sqrt{8} = i^2\sqrt{16} = -4.$

However, if we wrongly multiplied -2 by -8, we would get $+16 = +4$. Still not convinced? I will "wrongly" prove that $-1 = +1$.

Step	Reason
1. $-1 = -1$	Any number equals itself (known as the reflexive property).
2. $\frac{1}{-1} = \frac{-1}{1}$	A positive divided by a negative is the same as a negative divided by a positive.
3. $\sqrt{\frac{1}{-1}} = \sqrt{\frac{-1}{1}}$	Square roots of equals are equal.
4. $\frac{\sqrt{1}}{\sqrt{-1}} = \frac{\sqrt{-1}}{\sqrt{1}}$	Rule #4: $\sqrt{\frac{x}{y}} = \frac{\sqrt{x}}{\sqrt{y}}$.
5. $\frac{1}{i} = \frac{i}{1}$	$\sqrt{1} = 1$ and $\sqrt{-1} = i$.
6. $i \times \frac{1}{i} = i \times \frac{i}{1}$	If you multiply equals by equals, their products are equal.
7. $1 = -1$	$i^2 = -1$.

Did I just prove the impossible? Does 1 really equal -1? If you think it's true, you can give me one million dollars, but you still owe me one million since $1 = -1$.

Of course it's not true. See if you can figure out why. (See the end of the chapter if you can't.)

Example 16: Rewrite in terms of i:

Problem	Solution
a. $\sqrt{-9}$	$3i$
b. $-7 + \sqrt{-16}$	$-7 + 4i$
c. $10 + \sqrt{-32}$	$10 + 4i\sqrt{2}$

Chapter 4 Test

1. Simplify: $11\sqrt{84}$.
2. Simplify: $4x^2\sqrt{54x^7}$.
3. Simplify: $5\sqrt[3]{128}$.
4. Simplify: $4x\sqrt[4]{64x^{27}}$.
5. Simplify and combine: $4\sqrt{12} + 6\sqrt{18} + 8\sqrt{75}$.
6. Simplify and combine: $8z^4\sqrt{8z^5} - 10z^3\sqrt{32z^7}$.
7. $(4\sqrt{11})(7\sqrt{7}) =$
8. $4x\sqrt{8x} \times 5x^2\sqrt{3x^5} =$

Rationalize the denominator in questions 9 and 10:

9. $\dfrac{9a^9}{\sqrt{27a^{13}}} =$
10. $\dfrac{\sqrt{7}}{\sqrt{7} + \sqrt{3}} =$
11. If $5 + 6i = a + 3 + (b - 7)i$, a and $b =$
12. $i^{232} =$
13. $3(4i - 7) + 5i(3i - 2) =$
14. $i + i^2 + i^3 + i^4 =$
15. The magnitude of $-6 - 7i =$

Answers to Chapter 4 Test

1. $11 \times 2\sqrt{21} = 22\sqrt{21}$.
2. $(4x^2)(3x^3)\sqrt{6x} = 12x^5\sqrt{6x}$.
3. $5\sqrt[3]{64 \times 2} = 20\sqrt[3]{2}$.
4. $(4x)(2x^6)\sqrt[4]{4x^3} = 8x^7\sqrt[4]{4x^3}$.
5. $4 \times 2\sqrt{3} + 6 \times 3\sqrt{2} + 8 \times 5\sqrt{3} = 48\sqrt{3} + 18\sqrt{2}$.
6. $8z^4 \times 2z^2\sqrt{2z} - 10z^2 \times 4z^3\sqrt{2z} = -24z^6\sqrt{2z}$.
7. $28\sqrt{77}$.
8. $4x \times 2\sqrt{2x} \times 5x^2 \times x^2\sqrt{3x} = 40x^6\sqrt{6}$.
9. $\dfrac{9a^9 \times \sqrt{3a}}{3a^6\sqrt{3a} \times \sqrt{3a}} = a^2\sqrt{3a}$.
10. $\dfrac{\sqrt{7}\left(\sqrt{7}-\sqrt{3}\right)}{\left(\sqrt{7}+\sqrt{3}\right)\left(\sqrt{7}-\sqrt{3}\right)} = \dfrac{7-\sqrt{21}}{4}$.
11. $a + 3 = 5$ and $b - 7 = 6$; so $a = 2$ and $b = 13$.
12. $\dfrac{232}{4}$ has a remainder of 0; so the answer is the same as $i^0 = 1$.
13. $12i - 21 + 15i^2 - 10i = -21 - 15 + 2i = -36 + 2i$.
14. $i - 1 - i + 1 = 0$.
15. $|-6 - 7i| = \sqrt{(-6)^2 + (-7)^2} = \sqrt{85}$.

The answer to the wrong proof ($1 = -1$) is that you can't, can't, can't do step 4 since you have two square roots of negative numbers! Division of radicals works only for nonnegative numbers. It is illegal (a sick bird).

Now let's do some basic algebra that was taught in algebra 1 when I went to school several hundred years ago.

Answer to "Bob Asks": $B + D = 4$ and $B + 3D = 7$. Subtracting the second equation from the first, we get $2D = 3$. So a hot dog $D = \$1.50$. From the first equation, a burger $B = \$2.50$.

CHAPTER 5 *Vital Basics of Algebra*

"I like English puzzles also. Can you find a word with the letters WKW in it in that order?"

Algebraic manipulative skills such as those in this chapter are areas that high schools have tended to de-emphasize since 1985. This chapter is extremely important, both for the ACCUPLACER and success in college precalculus and calculus courses that you may need to take. If I needed a quick placement test for higher math courses, I might just give a test from this chapter only. As it turns out, the ACCUPLACER does have questions from this chapter. This is the chapter that you may need to study the most. You must know this chapter really well if you are going forward in math-related courses and want to succeed.

COMBINING LIKE TERMS

Like terms are terms with the same letter combination (or no letter). The same letter must also have the same exponents.

Example 1: Are the following terms like or unlike?

a. $4x$ and $-5x$

b. $4x$ and $4x^2$

c. xy^2 and x^2y

Solutions:

a. $4x$ and $-5x$ are like terms even though their numerical coefficients are different. We are concerned only with the same exponents.

b. $4x$ and $4x^2$ are unlike terms. They have different exponents.

c. xy^2 and x^2y are unlike; $xy^2 = xyy$ and $x^2y = xxy$. Again, different exponents.

Combining like terms means adding or subtracting their numerical coefficients; exponents are unchanged. **Unlike terms** cannot be combined.

Example 2: Simplify:

Problem	Solution
a. $3m + 4m + m =$	$8m$
b. $8m + 2n + 7m - 7n =$	$15m - 5n$
c. $3x^2 + 4x - 5 - 7x^2 - 4x + 8 =$	$-4x^2 + 3$

DISTRIBUTIVE LAW

The **distributive law** states:

$$a(x + y) = ax + ay$$

Example 3: Perform the indicated operations:

Problem	Solution
a. $4(3x - 7) =$	$12x - 28$
b. $5(2a - 5b + 3c) =$	$10a - 25b + 15c$
c. $3x^4(7x^3 - 4x - 1) =$	$21x^7 - 12x^5 - 3x^4$
d. $4(3x - 7) - 5(4x - 2) =$	$12x - 28 - 20x + 10 = -8x - 18$

BINOMIAL PRODUCTS

A **binomial** is a two-term expression, such as $x + 2$. We use the **FOIL method** to multiply a binomial by a binomial. FOIL is an acronym for First, Outer, Inner, Last. This means to multiply the first two terms, then the outer terms, then the inner terms, and finally the last two terms.

Example 4: Multiply $(x + 4)(x + 6)$

Solution:

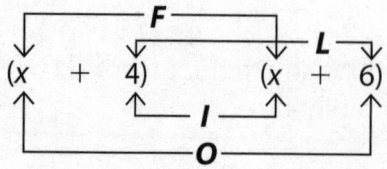

Multiplying, we get $x^2 + 6x + 4x + 24 = x^2 + 10x + 24$.

You should know the following common binomial products:

$$(a + b)(a - b) = a^2 - b^2$$

$$(a - b)(a - b) = a^2 - 2ab + b^2$$

$$(a + b)(a + b) = a^2 + 2ab + b^2$$

Note *For a perfect square $(a + b)^2$, the first term of the resulting trinomial is the first term squared (a^2), and the third term of the resulting trinomial is the last term squared (b^2). The middle term is twice the product of the two terms of the binomial ($2ab$), so*

$$(a + b)^2 = a^2 + 2ab + b^2$$

Example 5: Perform the indicated multiplications:

Problem	Solution
a. $(x + 7)(x + 4) =$	$x^2 + 4x + 7x + 28 = x^2 + 11x + 28$
b. $(x - 5)(x - 2) =$	$x^2 - 7x + 10$
c. $(x + 6)(x - 3) =$	$x^2 + 3x - 18$
d. $(x + 6)(x - 8) =$	$x^2 - 2x - 48$
e. $(x + 5)(x - 5) =$	$x^2 - 5x + 5x - 25 = x^2 - 25$
f. $(x + 5)^2 =$	$(x + 5)(x + 5) = x^2 + 10x + 25$
g. $(x - 10)^2 =$	$x^2 - 20x + 100$
h. $(2x + 5)(3x - 10) =$	$6x^2 - 5x - 50$
i. $3(x + 4)(x + 5) =$	$3(x^2 + 9x + 20) = 3x^2 + 27x + 60$
j. $7(4x + 3)(4x - 3) =$	$7(16x^2 - 9) = 112x^2 - 63$
k. $(a + 4)(a + 7)$	$a^2 + 11a + 28$
l. $(b + 5)(b + 6)$	$b^2 + 11b + 30$
m. $(c + 1)(c + 9)$	$c^2 + 10c + 9$

n. $(d + 4)(d + 8)$	$d^2 + 12d + 32$
o. $(e + 11)(e + 10)$	$e^2 + 21e + 110$
p. $(f - 6)(f - 2)$	$f^2 - 8f + 12$
q. $(g - 10)(g - 20)$	$g^2 - 30g + 200$
r. $(h - 4)(h - 3)$	$h^2 - 7h + 12$
s. $(j - 1)(j - 7)$	$j^2 - 8j + 7$
t. $(k - 3)(k - 5)$	$k^2 - 8k + 15$
u. $4\sqrt{2}(5\sqrt{7} + 6 + \sqrt{8} + 5\sqrt{14})$	$20\sqrt{14} + 24\sqrt{2} + 16 + 40\sqrt{7}$
v. $5x^2\sqrt{3x}(6x^4\sqrt{7x^6} + 4x\sqrt{3x})$	$30x^9\sqrt{21x} + 60x^4$
w. $(4\sqrt{2} + 6\sqrt{5})(8\sqrt{2} + 10\sqrt{5})$	$364 + 88\sqrt{10}$

Example 6: Perform the indicated multiplications:

Problem	Solution
a. $(k + 5)(k - 2)$	$k^2 + 3k - 10$
b. $(m + 5)(m - 8)$	$m^2 - 3m - 40$
c. $(n - 6)(n + 2)$	$n^2 - 4n - 12$
d. $(p - 8)(p + 10)$	$p^2 + 2p - 80$
e. $(q - 5r)(q + 2r)$	$q^2 - 3qr - 10r^2$
f. $(s + 3)^2$	$s^2 + 6s + 9$
g. $(t - 4)^2$	$t^2 - 8t + 16$
h. $(3u + 5)^2$	$9u^2 + 30u + 25$
i. $(5v - 4)^2$	$25v^2 - 40v + 16$

j. $(ax + by)^2$	$a^2x^2 + 2abxy + b^2y^2$
k. $(be - ma)^2$	$b^2e^2 - 2\,beam + m^2a^2$
l. $(w + x)(w - x)$	$w^2 - x^2$
m. $(a - 11)(a + 11)$	$a^2 - 121$
n. $(am - 7)(am + 7)$	$a^2m^2 - 49$
o. $(a^2b + c)(a^2b - c)$	$a^4b^2 - c^2$
p. $3(x + 5)(x - 2)$	$3x^2 + 9x - 30$
q. $-4(2x - 5)(3x - 4)$	$-24x^2 + 92x - 80$
r. $x(2x - 5)(4x + 7)$	$8x^3 - 6x^2 - 35x$
s. $5(x - 5)(x + 5)$	$5x^2 - 125$
t. $\left(a\sqrt{b} + c\sqrt{d}\right)\left(e\sqrt{f} + g\sqrt{h}\right)$	$ae\sqrt{bf} + ag\sqrt{bh} + ce\sqrt{df} + cg\sqrt{dh}$
u. $4x\sqrt{7x}\left(4x\sqrt{7x} + 3x^3\sqrt{5x^{11}}\right)$	$112x^3 + 12x^{10}\sqrt{35}$
v. $\left(2\sqrt{3} + 4\sqrt{5}\right)^2$	$92 + 16\sqrt{15}$
w. $\left(5\sqrt{u} + 3\sqrt{v}\right)^2$	$25u + 30\sqrt{uv} + 9v$
x. Rationalize: $\dfrac{\sqrt{y} - \sqrt{z}}{\sqrt{y} + \sqrt{z}}$	$\dfrac{y - 2\sqrt{yz} + z}{y - z}$

Q **Let's do a few multiple-choice exercises.**

Exercise 1: $x^2 - y^2 = 24$; $3(x + y)(x - y) =$

A. 8

B. 24

C. 27

D. 72

E. 13,824

Exercise 2: $x + y = m; x - y = \frac{1}{m}; x^2 - y^2 =$

A. m^2

B. m

C. 1

D. $\frac{1}{m}$

E. $\frac{1}{m^2}$

Exercise 3: $\left(x + \frac{1}{x}\right)^2 = 64; x^2 + \frac{1}{x^2} =$

A. 9

B. 62

C. 64

D. 65

E. 66

Exercise 4: $x^2 + y^2 = 20$ and $xy = -6$. Then $(x + y)^2 =$

A. 8

B. 14

C. 20

D. 26

E. 32

Exercise 5: $(x - y)^2 + 4xy =$

A. $x^2 + 8x + y^2$

B. $x^2 + 4x + y^2$

C. $x^2 - y^2$

D. $(x + y)^2$

E. $x^2 + y^2$

Let's look at the answers.

Answer 1: **D:** $3(x + y)(x - y) = 3(x^2 - y^2) = 3(24) = 72.$

Answer 2: **C:** $(x + y)(x - y) = x^2 - y^2 = \frac{m}{1} \times \frac{1}{m} = 1.$

Answer 3: **B:** $\left(x + \frac{1}{x}\right)\left(x + \frac{1}{x}\right) = x^2 + 2(x)\left(\frac{1}{x}\right) + \frac{1}{x^2} = x^2 + \frac{1}{x^2} + 2 = 64.$ So $x^2 + \frac{1}{x^2} = 64 - 2 = 62.$

Answer 4: **A:** $(x + y)^2 = x^2 + 2xy + y^2 = x^2 + y^2 + 2xy = 20 + 2(-6) = 8$.

Answer 5: **D:** $x^2 - 2xy + y^2 + 4xy = x^2 + 2xy + y^2 = (x + y)^2$.

Let's go on to factoring.

FACTORING

Factoring is the reverse of the distributive law. There are three types of factoring you need to know: largest common factor, difference of two squares, and trinomial factorization.

If the distributive law says $x(y + z) = xy + xz$, then taking out the largest common factor says $xy + xy = x(y + z)$. Let's demonstrate a few factoring examples.

Example 7: Factor:

Problem	Solution	Explanation
a. $4x + 6y - 8$	$2(2x + 3y - 4)$	2 is the largest common factor.
b. $8ax + 12ay - 40az$	$4a(2x + 3y - 10z)$	4 is the largest common factor; a is also a common factor.
c. $10a^4y^6z^3 - 15a^7y$	$5a^4y(2y^5z^3 - 3a^3)$	The largest common factor and the lowest power of each common variable is factored out: a^4 and y, but not z because it is not in both terms.
d. $x^4y - xy^3 + xy$	$xy(x^3 - y^2 + 1)$	Factor out the lowest power of each common variable. Three terms in the original give three terms in parentheses. Note that $1 \times xy = xy$.
e. $9by + 12be + 4ye$	prime	Some expressions cannot be factored.

Difference of Two Squares

Because $(a + b)(a - b) = a^2 - b^2$, factoring tells us that $a^2 - b^2 = (a + b)(a - b)$.

Example 8:

Problem	Solution	Explanation
a. $x^2 - 25$	$(x + 5)(x - 5)$ or $(x - 5)(x + 5)$	Either order is OK.
b. $x^2 - 121$	$(x + 11)(x - 11)$	
c. $9a^2 - 25b^2$	$(3a + 5b)(3a - 5b)$	
d. $5a^3 - 20a$	$5a(a^2 - 4) = 5a(a + 2)(a - 2)$	Factor out the largest common factor first, then use the difference of two squares.
e. $x^4 - y^4$	$(x^2 + y^2)(x^2 - y^2) =$ $(x^2 + y^2)(x + y)(x - y)$	This is the difference of two squares where the square roots in the factors are also squares. Sum of two squares doesn't factor, but use the difference of two squares again.

Example 9: Factor completely:

Problem	Solution
a. $12am + 18an$	$6a(2m + 3n)$
b. $6at - 18st + 4as$	$2(3at - 9st + 2as)$
c. $10ax + 15ae - 16ex$	Prime
d. $18a^5c^6 - 27a^3c^8$	$9a^3c^6(2a^2 - 3c^2)$
e. $25a^4b^7c^9 - 75a^8b^9c^{10}$	$25a^4b^7c^9(1 - 3a^4b^2c)$
f. $a^4b^5 + a^7b - ab$	$ab(a^3b^4 + a^6 - 1)$
g. $9 - x^2$	$(3 + x)(3 - x)$

h. $x^4 - 36y^2$	$(x^2 + 6y)(x^2 - 6y)$
i. $2x^3 - 98x$	$2x(x + 7)(x - 7)$
j. $a^4 - 81b^2$	$(a^2 + 9b)(a^2 - 9b)$
k. $x^2 - 49$	$(x + 7)(x - 7)$
l. $5z^2 - 25$	$5(z^2 - 5)$
m. $a^4 - c^8$	$(a^2 + c^4)(a + c^2)(a - c^2)$
n. $2a^9 - 32a$	$2a(a^4 + 4)(a^2 + 2)(a^2 - 2)$

Q **Let's do a couple more multiple-choice exercises.**

Exercise 6: $8x + 6y = 30$. Then $20x + 15y =$

A. 15
B. 30
C. 60
D. 75
E. 150

Exercise 7: $x^2 - 4 = 47 \times 43$; $x =$

A. 41
B. 43
C. 45
D. 47
E. 49

A **Let's look at the answers.**

Answer 6: D: $8x + 6y = 2(4x + 3y) = 30$; so $4x + 3y = 15$. $20x + 15y = 5(4x + 3y) = 5 \times 15 = 75$.

Answer 7: C: $x^2 - 4 = (x + 2)(x - 2) = 47 \times 43$, or $(45 + 2)(45 - 2)$. So $x = 45$.

Later in this chapter as well as in later chapters, we will see more comparison questions. For now, let's do trinomial factoring.

Factoring Trinomials

Factoring trinomials is a puzzle, a game, which is rarely done well in high school and even more rarely practiced. Let's learn the factoring game.

First, let's rewrite the first four parts of Example 5 backward and look at them.

a. $x^2 + 11x + 28 = (x + 7)(x + 4)$

b. $x^2 - 7x + 10 = (x - 5)(x - 2)$

c. $x^2 + 3x - 18 = (x + 6)(x - 3)$

d. $x^2 - 2x - 48 = (x + 6)(x - 8)$

Each term starts with x^2 ($= +1x^2$), so the first sign is $+$. We'll call the sign in front of the x term the middle sign, and we'll call the sign in front of the number term the last sign.

Let's look at **a** and **b** above to state some rules of the game:

1. If the last sign (in the trinomial) is $+$, then both signs (in the parentheses) must be the same. The reason? $(+) \times (+) = +$, and $(-) \times (-) = +$.
2. Only if the last sign is $+$, look at the sign of the middle term. If it is $+$, both factors have a $+$ sign (as in **a**); if it is $-$, both factors have a $-$ sign (as in **b**).
3. If the last sign is $-$, the signs in the two factors must be different (see **c** and **d**).

Now, let's play the game:

Example 10: Factor: $x^2 - 16x + 15$.

Solution:

1. The last sign $+$ means both signs are the same. The middle sign $-$ means both are $-$.
2. The only factors of x^2 are $(x)(x)$. Look at the number term 15. The factors of 15 are (3)(5) and (1)(15). So $(x - 5)(x - 3)$ and $(x - 15)(x - 1)$ are the only possibilities. We have chosen the first and last terms to be correct, so we need to check only the middle term. The first, $-8x$, is wrong; the second, $-16x$, is correct.
3. The answer is $(x - 15)(x - 1)$. If neither worked, the trinomial couldn't be factored.

Example 11: Factor completely: $x^2 - 4x - 21$.

Solution:

1. The last sign negative means the signs of the factors are different.
2. $x^2 = x(x)$ and the factors of 21 are 7 and 3 or 21 and 1. We want the pair that totals the middle number (-4), so -7 and 3 are correct.

Note that the larger factor (7) gets the minus sign because the middle number is negative.

3. The answer is $(x - 7)(x + 3)$.

Note *If you multiply the inner and outer terms and get the right number but the wrong sign, both signs in the parentheses must be changed.*

The game gets more complicated if the coefficient of x^2 is not 1.

Example 12: Factor completely: $4x^2 + 4x - 15$.

Solution:

1. The last sign is $-$, so the signs of the factors must be different.
2. The factors of $4x^2$ are $(4x)(x)$ or $(2x)(2x)$.
3. The factors of 15 are 3 and 5, or 1 and 15. Let's write out all the possibilities and the resulting middle terms. We are looking for terms whose difference is $4x$.

   $(4x \quad 3)(x \quad 5)$; middle terms are $3x$ and $20x$, no way to get $4x$.

   $(4x \quad 5)(x \quad 3)$; middle terms are $5x$ and $12x$, and again, there is no way to get $4x$.

   $(4x \quad 15)(x \quad 1)$; middle terms are $15x$ and $4x$, wrong!

   $(4x \quad 1)(x \quad 15)$; middle terms are $1x$ and $60x$, the next county.

   $(2x \quad 1)(2x \quad 15)$; middle terms are $2x$ and $30x$, wrong again!

   $(2x \quad 3)\ (2x \quad 5)$; middle terms are $6x$ and $10x$, correct, whew!
4. The minus sign goes in front of the 3 and the plus sign goes in front of the 5 to get the middle coefficient of $+4$, so the answer is $(2x - 3)(2x + 5)$.

Example 13: Factor completely: $3x^2 + 15x + 12$.

Solution: Take out the common factor first.

$$3x^2 + 15x + 12 = 3(x^2 + 5x + 4) = 3(x + 4)(x + 1).$$

Example 14. Factor completely; coefficients may be integers only.

Problem	Solution
a. $x^2 + 11x + 24$	$(x + 3)(x + 8)$
b. $x^2 - 11x - 12$	$(x - 12)(x + 1)$
c. $x^2 + 5x - 6$	$(x + 6)(x - 1)$

d. $x^2 - 20x + 100$	$(x - 10)^2$
e. $x^2 - x - 2$	$(x - 2)(x + 1)$
f. $x^2 - 15x + 56$	$(x - 7)(x - 8)$
g. $x^2 + 8x + 16$	$(x + 4)^2$
h. $x^2 - 6x - 16$	$(x - 8)(x + 2)$
i. $x^2 - 17x + 42$	$(x - 14)(x - 3)$
j. $x^2 + 5xy + 6y^2$	$(x + 2y)(x + 3y)$
k. $3x^2 - 6x - 9$	$3(x - 3)(x + 1)$
l. $4x^2 + 16x - 20$	$4(x + 5)(x - 1)$
m. $x^3 - 12x^2 + 35x$	$x(x - 7)(x - 5)$
n. $2x^8 + 8x^7 + 6x^6$	$2x^6(x + 3)(x + 1)$
o. $x^4 - 10x^2 + 9$	$(x + 3)(x - 3)(x + 1)(x - 1)$
p. $x^4 - 8x^2 - 9$	$(x^2 + 1)(x + 3)(x - 3)$
q. $2x^2 - 5x + 3$	$(2x - 3)(x - 1)$
r. $2x^2 + 5x - 3$	$(2x - 1)(x + 3)$
s. $5x^2 - 11x + 2$	$(5x - 1)(x - 2)$
t. $9x^2 + 21x - 8$	$(3x + 8)(3x - 1)$
u. $3x^2 - 8x - 3$	$(3x + 1)(x - 3)$
v. $6x^2 - 13x + 6$	$(3x - 2)(2x - 3)$
w. $6x^2 + 35x - 6$	$(6x - 1)(x + 6)$
x. $9x^2 + 71x - 8$	$(9x - 1)(x + 8)$
y. $9x^4 + 24x^3 + 12x^2$	$3x^2(3x + 2)(x + 2)$

Q Let's do a couple of multiple-choice exercises.

Exercise 8: If $x^2 + 5x + 4 = 27$, then $3(x + 1)(x + 4) =$

A. 9
B. 24
C. 30
D. 81
E. 243

Exercise 9: If $(x - 6)$ is a factor of $x^2 + kx - 48$, $k =$

A. -288
B. -14
C. -2
D. 2
E. 14

A Let's look at the answers.

Answer 8: **D:** $x^2 + 5x + 4 = (x + 1)(x + 4) = 27$; $3(x + 1)(x + 4) = 3(27) = 81$.

Answer 9: **D:** $(-6)(?) = -48$; $? = 8$; $8 + (-6) = 2 = k$. To check, $(x - 6)(x + 8) = x^2 + 2x - 48$.

Again, later on, we will present more problems involving trinomial factoring.

There are two other factorings that may occur. They are less important. However, many schools require you to know one or both of them.

Sum and Differences of Two Cubes

We already saw that the difference of two squares factors into real numbers. For example, $x^2 - 9 = (x - 3)(x + 3)$.

The sum of two squares, such as $x^2 + 9$, does not factor into real numbers. However, with cubes, it is a different story.

$$x^3 - y^3 = (x - y)(x^2 + xy + y^2)$$

and

$$x^3 + y^3 = (x + y)(x^2 - xy + y^2).$$

Note *The proof would be to multiply out the right sides and get the left side.*

Note *FYI: $x^2 + 9 = (x + 3i)(x - 3i)$. We will not do this kind of factoring in this book.*

Example 15: Factor:

Problem	Solution
a. $x^3 - 27$	$(x)^3 - (3)^3 = (x - 3)(x^2 + 3x + 9)$
b. $y^6 + 64$	$(y^2)^3 + (4)^3 = (y^2 + 4)(y^4 - 4y^2 + 16)$
c. $x^6 - y^6$	$(x^3 + y^3)(x^3 - y^3) = (x + y)$
	$(x^2 - xy + y^2)(x - y)(x^2 + xy + y^2)$

Grouping

For an expression such as $am + em + ah + eh$:

1. We first see there is no common factor.
2. There is no difference of two squares since there are no squares and there are more than two terms.
3. There is no difference or sum of two cubes since there are no cubes and there are more than two terms.
4. There is no trinomial factoring since there are more than three terms.

Grouping the terms sometimes allows us to factor an expression. We notice in this case that an m can be factored from the first two terms, and an h can be factored from the last two. So $am + em + ah + eh = m(a + e) + h(a + e)$. Now we can see that the $(a + e)$ is a common factor! So the final answer is $(a + e)(m + h)$.

Note *Always try the first four types of factoring in that order. However, if grouping is possible, you may have to then use one of the first four steps.*

Example 16: Factor $x^3 - x^2 - x + 1$.

Solution: $x^3 - x^2 - x + 1 = x^2(x - 1) - 1(x - 1)$. We have to take out $(x - 1)$ from the second two terms to make both parentheses the same! When we take out the $(x - 1)$, we get $(x - 1)(x^2 - 1) = (x + 1)(x - 1)(x - 1)$, or $(x + 1)(x - 1)^2$.

Other groupings are also possible, as shown in the following example.

Example 17: Factor $x^2 - 4x + 4 - y^2$.

Solution: We group this expression into three and one terms instead of two and two: $(x^2 - 4x + 4) - y^2 = (x - 2)(x - 2) - y^2 = (x - 2)^2 - y^2 = (x - 2 + y)(x - 2 - y)$, by using the difference of two squares.

With grouping, we could go on almost forever, but we won't. This is enough!

Example 18: Factor the following expressions:

Problem	Solution
a. $a^3 - 125$	$(a - 5)(a^2 + 5a + 25)$
b. $8b^3 + 343$	$(2b + 7)(4b^2 - 14b + 49)$
c. $c^6 + 27$	$(c^2 + 3)(c^4 - 3c^2 + 9)$
d. $d^6 - 1{,}000$	$(d^2 - 10)(d^4 + 10d^2 + 100)$
e. $8e^3 - 64$	$8(e^3 - 8) = 8(e - 2)(e^2 + 2e + 4)$
f. $f^{12} - z^{12}$	$(f^6 + z^6)(f^6 - z^6) =$
	$(f^2 + z^2)(f^4 - f^2z^2 + z^4)(f^3 + z^3)(f^3 - z^3) =$
	$(f^2 + z^2)(f^4 - f^2z^2 + z^4)$
	$(f + z)(f^2 - fz + z^2) \times$
	$(f - z)(f^2 + fz + z^2)$
g. $gx + gy + wx + wy$	$g(x + y) + w(x + y) = (g + w)(x + y)$
h. $x^3 + 3x^2 + 3x + 1$	$x^3 + 1 + 3x^2 + 3x =$
	$(x + 1)(x^2 - x + 1) + 3x(x + 1) =$
	$(x + 1)(x^2 - x + 1 + 3x) =$
	$(x + 1)(x^2 + 2x + 1) =$
	$(x + 1)(x + 1)(x + 1) = (x + 1)^3$

i. $x^2 - 6xy + 9y^2 - z^2$ $\quad (x - 3y)^2 - z^2 = (x - 3y + z)(x - 3y - z)$

j. $9 - a^2 - 2ab - b^2$ $\quad 9 - (a + b)^2 = [3 - (a + b)][3 + (a + b)] =$

$(3 - a - b)(3 + a + b)$

k. $3x - 3y + x^2 - 2xy + y^2$ $\quad 3(x - y) + (x - y)^2 = (x - y)(3 + x - y)$

(a 2 and 3 grouping)

I just put the last few in for those who want a little more challenge. If you want a super challenge, try to factor $x^4 + x^2 + 1$. Yes, it can factor. The answer is at the end of the chapter.

ALGEBRAIC FRACTIONS

Except for adding and subtracting, the techniques for algebraic fractions are easy to understand. They must be practiced, however.

Reducing Fractions

Factor the top and bottom; cancel factors that are the same.

Example 19: Reduce the following fractions:

Problem	Solution
a. $\dfrac{x^2 - 9}{x^2 - 3x}$	$\dfrac{(x + 3)(x - 3)}{x(x - 3)} = \dfrac{x + 3}{x}$
b. $\dfrac{2x^3 + 10x^2 + 8x}{x^4 + x^3}$	$\dfrac{2x(x + 4)(x + 1)}{x^3(x + 1)} = \dfrac{2(x + 4)}{x^2}$
c. $\dfrac{x - 9}{9 - x}$	$\dfrac{(x - 9)}{-1(x - 9)} = -1$

Let's do a few more multiple-choice exercises.

Exercise 10: $\dfrac{16x + 4}{4} =$

A. $4x$

B. $4x + 1$

C. $12x$

D. $12x + 1$

E. $16x + 1$

Exercise 11: $\dfrac{x^2 - 16}{x - 4} =$

A. $x + 4$

B. $x - 4$

C. $2x + 4$

D. $2x - 4$

E. $x^2 - x - 12$

Exercise 12: $\dfrac{2x^2 + 10x + 12}{2x + 6} =$

A. x

B. $x + 1$

C. $x + 2$

D. $11x + 2$

E. $11x + 6$

Let's look at the answers.

Answer 10: B: $\dfrac{16x + 4}{4} = \dfrac{4(4x + 1)}{4} = 4x + 1$. Another way to do this problem is

$$\frac{16x + 4}{4} = \frac{16x}{4} + \frac{4}{4} = 4x + 1.$$

Answer 11: A: $\dfrac{(x + 4)(x - 4)}{x - 4} = x + 4.$

Answer 12: C: $\dfrac{2(x + 2)(x + 3)}{2(x + 3)} = \dfrac{2(x + 2)}{2} = x + 2.$

Multiplication and Division of Fractions

Algebraic fractions use the same principles as multiplication and division of numerical fractions except we factor all tops and bottoms, canceling one factor in any top with its equivalent in any bottom. In a division problem, we must remember to first invert the second fraction and then multiply.

Example 20: $\dfrac{x^2 - 25}{(x + 5)^3} \times \dfrac{x^3 + x^2}{x^2 - 4x - 5} =$

Solution: $\dfrac{(x + 5)(x - 5)}{(x + 5)(x + 5)(x + 5)} \times \dfrac{x^2(x + 1)}{(x - 5)(x + 1)} = \dfrac{x^2}{(x + 5)^2}$

Example 21: $\dfrac{x^4 + 4x^2}{x^6} \div \dfrac{x^4 - 16}{x^2 + 3x - 10} =$

Solution: $\dfrac{x^2(x^2 + 4)}{x^6} \times \dfrac{(x + 5)(x - 2)}{(x^2 + 4)(x - 2)(x + 2)} = \dfrac{x + 5}{x^4(x + 2)}$

Adding and Subtracting Algebraic Fractions

It might be time to review the section in Chapter 2 on adding and subtracting fractions. Follow these steps:

1. If the bottoms (denominators) are the same, add (or subtract) the tops (numerators), reducing if necessary.
2. If the bottoms are different:
   a. Factor the denominators of each term.
   b. Determine the least common denominator (LCD), the product of the most number of times a prime appears in any one denominator.
   c. Multiply top and bottom of each term by "what's missing" (i.e., the LCD divided by the denominator of that term).
   d. Add (or subtract) and simplify the numerators; reduce if possible.

Note *The LCD is simply the least common multiple (LCM) of the denominators. LCM is discussed in Chapter 2.*

Example 22: $\dfrac{x}{36 - x^2} - \dfrac{6}{36 - x^2} =$

Solution: $\dfrac{x - 6}{36 - x^2} = \dfrac{x - 6}{(6 - x)(6 + x)} = \dfrac{-1}{x + 6}$

Example 23: $\dfrac{5}{12xy^3} + \dfrac{9}{8x^2y} =$

Solution: $\dfrac{5}{2 \times 2 \times 3xyyy} + \dfrac{9}{2 \times 2 \times 2xxy}$

$= \dfrac{5(2x)}{2 \times 2 \times 2 \times 3xxyyy} + \dfrac{9(3yy)}{2 \times 2 \times 2 \times 3xxyyy} = \dfrac{10x + 27y^2}{24x^2y^3}$

Example 24: $\dfrac{2}{x^2 + 4x + 4} + \dfrac{3}{x^2 + 5x + 6} =$

Solution: $\dfrac{2}{(x + 2)(x + 2)} + \dfrac{3}{(x + 2)(x + 3)}$

$$= \frac{2(x + 3)}{(x + 2)(x + 2)(x + 3)} + \frac{3(x + 2)}{(x + 2)(x + 3)(x + 2)} = \frac{5x + 12}{(x + 2)(x + 2)(x + 3)}$$

Simplifying Complex Fractions

The topic of complex fractions (fractions with fractions in the numerator and/or denominator) is neglected in almost every algebra course. You need to be able to simplify complex fractions on the ACCUPLACER, with both numbers and letters. To do so, you simply multiply all terms by the LCD of the fractions.

Example 25: Simplify $\dfrac{2 - \frac{5}{6}}{\frac{2}{3} + \frac{7}{8}}$.

Solution: Find the LCD of all the terms (24) and multiply each term by it.

$$\frac{\frac{2}{1} \times \frac{24}{1} - \frac{5}{6} \times \frac{24}{1}}{\frac{2}{3} \times \frac{24}{1} + \frac{7}{8} \times \frac{24}{1}} = \frac{48 - 20}{16 + 21} = \frac{28}{37}$$

Note *When you multiply each term by the LCD, all the fractions disappear except for the major one.*

Example 26: Simplify $\dfrac{1 - \frac{25}{x^2}}{1 - \frac{10}{x} + \frac{25}{x^2}}$.

Solution: The LCD is x^2.

$$\frac{x^2\left(1 - \frac{25}{x^2}\right)}{x^2\left(1 - \frac{10}{x} + \frac{25}{x^2}\right)} = \frac{x^2 - 25}{x^2 - 10x + 25} = \frac{(x + 5)(x - 5)}{(x - 5)(x - 5)} = \frac{x + 5}{x - 5}$$

Example 27: Simplify the following expressions:

Problem	Solution
a. $\dfrac{x^2 - 4x + 4}{x^2 - 2x}$	$\dfrac{x - 2}{x}$
b. $\dfrac{(x + 5)^5}{x^3 + 10x^2 + 25x}$	$\dfrac{(x + 5)^3}{x}$

c. $\dfrac{a+b-c+d-e}{-a-b+c-d+e}$ $\qquad -1$

d. $\dfrac{x^2-9}{x^2-9x} \times \dfrac{x^8}{x^2+4x+3}$ $\qquad \dfrac{x^7(x-3)}{(x-9)(x+1)}$

e. $\dfrac{4x-8}{6x^2-6x-12} \times \dfrac{x^2+2x+1}{2x^2+5x+3}$ $\qquad \dfrac{2}{3(2x+3)}$

f. $\dfrac{x^2-y^2}{x^2-2xy+y^2} \times \dfrac{x^2-4xy+3y^2}{x^2-2xy-3y^2}$ $\qquad 1$

g. $\dfrac{4x^2-9}{2x^2-x-3} \div \dfrac{4x^2+12x+9}{x^2+x}$ $\qquad \dfrac{x}{2x+3}$

h. $\dfrac{x^3-27}{x^3+1} \div \dfrac{2x^3+6x^2+18x}{x^3-x^2-x+1}$ $\qquad \dfrac{(x-3)(x-1)^2}{2x(x^2-x+1)}$

i. $\dfrac{x}{9-x^2} - \dfrac{3}{9-x^2}$ $\qquad \dfrac{-1}{3+x}$

j. $\dfrac{x}{x^2-16} + \dfrac{4}{x+4}$ $\qquad \dfrac{5x-16}{(x+4)(x-4)}$

k. $\dfrac{3x}{x-6} - \dfrac{4}{x+2}$ $\qquad \dfrac{3x^2+2x+24}{(x-6)(x+2)}$

l. $\dfrac{2}{x^2+5x+6} + \dfrac{3}{x^2+6x+9} + \dfrac{4}{x^2+7x+12}$ $\qquad \dfrac{9x^2+52x+72}{(x+2)(x+3)(x+3)(x+4)}$

m. $\dfrac{2-\frac{11}{3}}{\frac{7}{2}-\frac{1}{6}}$ $\qquad -\dfrac{1}{2}$

n. $\dfrac{\frac{1}{x}}{\frac{1}{x^2}-1}$ $\qquad \dfrac{-x}{(x+1)(x-1)}$

o. $\dfrac{\frac{1}{x^2}-\frac{1}{4}}{\frac{1}{x}+\frac{1}{2}}$ $\qquad \dfrac{2-x}{2x}$

Q Let's do some multiple-choice exercises.

Exercise 13: The reciprocal of $2 - \frac{1}{4}$ is

A. $\frac{1}{2} - 4$

B. $-\frac{4}{7}$

C. $-\frac{4}{9}$

D. $\frac{4}{9}$

E. $\frac{4}{7}$

Exercise 14: $y = \frac{1}{x}$. Then $\frac{1-x}{1-y} =$

A. $-x$

B. x

C. 1

D. $(1 - x)$

E. $(x - 1)$

Exercise 15: $\frac{2\frac{1}{5} + \frac{3}{5}}{2\frac{1}{5} - \frac{3}{5}} =$

A. -1

B. 1

C. $\frac{7}{5}$

D. $\frac{7}{4}$

E. $\frac{14}{5}$

A Let's look at the answers.

Answer 13: E: $2 - \frac{1}{4} = \frac{7}{4}$. Its reciprocal is $\frac{4}{7}$.

Answer 14: A: $\frac{1-x}{1-y} = \frac{1-x}{1-\frac{1}{x}} = \frac{x(1-x)}{x(1-\frac{1}{x})} = \frac{x(1-x)}{x-1} = -x$. The answer is A because $\frac{1-x}{x-1} = -1$.

Answer 15: D: The top is $\frac{11}{5} + \frac{3}{5} = \frac{14}{5}$; the bottom is $\frac{11}{5} - \frac{3}{5} = \frac{8}{5}$. So the fraction simplifies to $\frac{\frac{14}{5}}{\frac{8}{5}}$; the 5's cancel, and $\frac{14}{8} = \frac{7}{4}$.

SHORT AND LONG DIVISION

Algebraic **short division** is the opposite of adding fractions with like denominators. In symbols, for example, $\frac{a+b-c}{d} = \frac{a}{d} + \frac{b}{d} - \frac{c}{d}$, where d is a single term, a monomial.

Example 28:

	Problem	Solution
a.	$\frac{12x-18}{6}$	$2x-3$
b.	$\frac{8x^{10}-12x^8-20x^7}{4x^4}$	$2x^6-3x^4-5x^3$
c.	$\frac{abc-abd-ab}{ab}$	$c-d-1$

For completeness, we will do algebraic long division. First, let's review long division with numbers.

Suppose we have $2371 \div 6$. How would we do this problem?

1. Divide 6 into 23 and get 3.
2. $3 \times 6 = 18$.
3. $23 - 18 = 5$.
4. Bring down the 7 to get 57. Repeat.
5. Divide 6 into 57 and get 9.
6. $9 \times 6 = 54$.
7. $57 - 54 = 3$.
8. Bring down the 1 to get 31. Repeat.
9. Divide 6 into 31 and get 5.
10. $5 \times 6 = 30$.
11. $31 - 30 = 1$. The remainder is 1.
12. The answer is $395\frac{1}{6}$.

$$
\begin{array}{r}
395\ \tfrac{1}{6} \\
6\overline{)2371} \\
\underline{18} \\
57 \\
\underline{54} \\
31 \\
\underline{30} \\
1
\end{array}
$$

Note *The 6 is called the **divisor**; 2371 is called the **dividend**; $395\frac{1}{6}$ is called the **quotient**.*

The procedure for **algebraic long division** is the same.

Example 29: Divide $4x^3 - 2x^2 + 8x - 11$ by $2x - 4$

Solution:

$$
\begin{array}{r}
2x^2 + 3x + 10 + \dfrac{29}{2x-4} \quad x \neq 2 \\
2x-4\overline{)\,4x^3 - 2x^2 + 8x - 11} \\
\underline{\mp 4x^3 \pm 8x^2} \\
6x^2 + 8x \\
\underline{\mp 6x^2 \pm 12x} \\
20x - 11 \\
\underline{\mp 20x \pm 40} \\
29
\end{array}
$$

1. If the divisor and/or the dividend are not arranged from the highest exponent to the lowest, rearrange the terms. In this problem, they are already arranged highest to lowest.
2. Divide the highest power term of the divisor into the highest power term of the dividend. $\frac{4x^3}{2x} = 2x^2$.
3. Multiply $2x^2$ by the entire divisor: $2x^2(2x - 4) = 4x^3 - 8x^2$.
4. Subtract: $4x^3 - 2x^2 + 80 - 11 - (4x^3 - 8x^2) = 6x^2 + 8x - 11$. Repeat.

Note *We have already brought down the next term.*

Note *The first terms must cancel; otherwise, there is a mistake.*

5. $\frac{6x^2}{2x} = 3x$; $3x(2x - 4) = 6x^2 - 12x$; $6x^2 + 8x - 11 - (6x^2 - 12x)$ $= 20x - 11$. Repeat.
6. $\frac{20x}{2x} = 10$; $10(2x - 4) = 20x - 40$; $20x - 11 - (20x - 40) = 29$; the remainder is 29.
7. The final answer is $2x^2 + 3x + 10 + \frac{29}{2x - 4}$; $x \neq 2$ because you cannot divide by 0.

Note *If the remainder had been 0 in this problem, that would have meant $2x - 4$ was a factor of the original polynomial.*

Example 30: Perform the following divisions:

	Problem	Solution
a.	$\frac{20x + 5}{5}$	$4x + 1$
b.	$\frac{20x^8 - 12x^5 - 4x^2}{4x^2}$	$5x^6 - 3x^3 - 1$
c.	$\frac{12x^6y^8 + 10x^9y^3}{4x^2y^2}$	$3x^4y^6 + \frac{5}{2}x^7y$ or $\frac{6x^4y^6 + 5x^7y}{2}$
d.	$\frac{4x^3 - 3x^2 + 2x - 1}{x + 1}$	$4x^2 - 7x + 9 - \frac{10}{x + 1}$
e.	$\frac{4x^3 - 6x^2 + 8x - 10}{2x - 4}$	$2x^2 + x + 6 + \frac{14}{2x - 4}$
f.	$\frac{x^3 + y^3}{x + y}$	$x^2 - xy + y^2$

Chapter 5 Test

1. Simplify: $4x^2 - 5y^3 - 6x^2 - 7y^3$.
2. Multiply and simplify: $4x(2x - 6) - 3x(8x - 1)$.
3. Multiply: $(3x - 4)(5x + 7)$.
4. Multiply: $4\sqrt{7}\left(3\sqrt{7} - 4 - 6\sqrt{14}\right)$.
5. $\left(x^3 - 3\sqrt{y}\right)^2 =$
6. Factor completely: $4x^4 - 64$.
7. Factor completely: $9x^2 + 21x - 8$.
8. Factor completely: $c^6 - d^6$.
9. Reduce: $\dfrac{4 - x^2}{x^2 - 4x + 4}$.
10. Add: $\dfrac{5}{x - 5} + \dfrac{2}{x^2 - 6x + 5}$.
11. Subtract: $\dfrac{4x}{x + 2} - \dfrac{2x}{x + 4}$.
12. Multiply: $\dfrac{2x + 5}{4x^2 - 25} \times \dfrac{2x - 5}{4x^2 - 20x + 25}$.
13. Divide: $\dfrac{6x - 9}{10x - 15} \div (x - 4)$.
14. Divide: $\dfrac{x^2 - 25}{x^2 - x - 20} \div \dfrac{x^2 + 8x + 15}{x^2 + 7x + 12}$.
15. Divide: $\dfrac{2x^3 - 2x^2 - 2x - 1}{2x + 2}$.
16. Divide: $\dfrac{10x^{10} - 8x^8}{2x^2}$.
17. Compute without a calculator: $\dfrac{\frac{1}{2} + \frac{2}{3}}{\frac{5}{6} + \frac{7}{8}} =$
18. Simplify: $\dfrac{\frac{1}{x^2} - \frac{1}{y^2}}{\frac{1}{x} - \frac{1}{y}}$.
19. Simplify: $\dfrac{6x - 4}{2} - \dfrac{15x - 12}{3}$.
20. Simplify: $\dfrac{x^2 - 4x - 5}{x^2 - 7x + 10} - \dfrac{x^2 - 5x + 6}{x^2 - 2x - 3}$.

21. Rationalize the denominator: $\dfrac{\sqrt{z}+\sqrt{5}}{\sqrt{z}-\sqrt{5}}$.

22. $(3x-7)(4x-2)-(2x-6)(6x+1)=$

23. Evaluate (answer in terms of i): $(4-\sqrt{-9})(2+\sqrt{-4})$.

24. Subtract: $2-\dfrac{x-6}{x-3}$.

25. Evaluate (answer in terms of i): $\dfrac{4+3i}{3-7i}$.

Answers to Chapter 5 Test

1. $-2x^2-12y^3$.
2. $-16x^2-21x$.
3. $15x^2+x-28$.
4. $84-16\sqrt{7}-168\sqrt{2}$.
5. $x^6-6x^3\sqrt{y}+9y$.
6. $4(x^2+4)(x+2)(x-2)$.
7. $(3x+8)(3x-1)$.
8. $(c^3-d^3)(c^3+d^3)=(c-d)(c^2+cd+d^2)(c+d)(c^2-cd+d^2)$.
9. $\dfrac{(2-x)(2+x)}{(x-2)(x-2)}=\dfrac{-(x+2)}{(x-2)}$ or $\dfrac{x+2}{2-x}$.
10. $\dfrac{5(x-1)}{(x-5)(x-1)}+\dfrac{2}{(x-5)(x-1)}=\dfrac{5x-3}{(x-5)(x-1)}$.
11. $\dfrac{4x(x+4)-2x(x+2)}{(x+2)(x+4)}=\dfrac{2x^2+12x}{(x+2)(x+4)}$.
12. $\dfrac{(2x+5)}{(2x+5)(2x-5)}\times\dfrac{(2x-5)}{(2x-5)\,(2x-5)}=\dfrac{1}{(2x-5)\,(2x-5)}$.
13. $\dfrac{3(2x-3)}{5(2x-3)}\times\dfrac{1}{(x-4)}=\dfrac{3}{5(x-4)}$.
14. $\dfrac{(x-5)(x+5)}{(x-5)(x+4)}\times\dfrac{(x+4)(x+3)}{(x+3)(x+5)}=1$.
15. $x^2-2x+1+\dfrac{-3}{2x+2}$.
16. $5x^8-4x^6$.

17. Multiply all fractions by the LCD, 24: $\frac{12+16}{20+21} = \frac{28}{41}$.

18. Multiply all fractions by x^2y^2, factor, and reduce: $\frac{y^2 - x^2}{xy^2 - yx^2} = \frac{y+x}{xy}$.

19. $-2x + 2$.

20. $\frac{(x-5)(x+1)}{(x-5)(x-2)} - \frac{(x-2)(x-3)}{(x+1)(x-3)} = \frac{(x+1)(x+1) - (x-2)(x-2)}{(x-2)(x+1)} = \frac{6x-3}{(x-2)(x+1)}$.

21. $\frac{(\sqrt{z}+\sqrt{5})(\sqrt{z}+\sqrt{5})}{(\sqrt{z}-\sqrt{5})(\sqrt{z}+\sqrt{5})} = \frac{z+2\sqrt{5z}+5}{z-5}$.

22. $12x^2 - 34x + 14 - (12x^2 - 34x - 6) = 20$.

23. $(4 - 3i)(2 + 2i) = 8 + 8i - 6i - 6i^2 = 14 + 2i$.

24. $\frac{2(x-3)}{(x-3)} - \frac{x-6}{x-3} = \frac{x}{x-3}$.

25. $\frac{(4+3i)(3+7i)}{(3-7i)(3+7i)} = \frac{-9+37i}{58} = \frac{-9}{58} + \frac{37}{58}i$.

Here is the solution to the factoring of $x^4 + x^2 + 1$. The only factors of x^4 are x^2 and x^2. The only factors of 1 are 1 and 1. All signs are plus. But if we multiply $(x^2 + 1)(x^2 + 1)$, we get $x^4 + 2x^2 + 1$. So we have an extra x^2, which has to be subtracted. So $x^4 + x^2 + 1 = (x^2 + 1)(x^2 + 1) - x^2$, or $(x^2 + 1)^2 - x^2$, which is the difference of two squares!!!!! Then the answer is $(x^2 + 1 + x)(x^2 + 1 - x)$. For me, there are always more and more interesting problems in math (and English also). Perhaps for you too?

Let's go on to equations.

Answer to "Bob Asks": Awkward.

CHAPTER 6: *Equations and Inequalities*

"Can you find a word with the letters HYX in it, in that order?"

FIRST-DEGREE EQUATIONS

In high school, the topic of first-degree equations was probably the most popular of all. The ACCUPLACER asks questions that are usually not too long and usually not too tricky. To review, here are the steps to solving first-degree equations. If you get good at these, you will know when to use the steps in another order.

To solve for *x*, follow these steps (believe it or not, it took a long time to get the phrasing of this list just right):

1. Multiply by the LCD to get rid of fractions. Cross-multiply if there are only two fractions.
2. If the "*x*" term appears only on the right, switch the sides.
3. Multiply out all parentheses by using the distributive law.
4. On each side, combine like terms.
5. Add the opposite of the *x* term on the right to each side.
6. Add the opposite of the non-*x* term(s) on the left to each side.
7. Factor out the *x*. This step occurs only if there is more than one letter in a problem.
8. Divide each side by the whole coefficient of *x*, including the sign.

Note ***Cross-multiplication*** *works when two fractions are equal. If you multiply the numerator of one by the denominator of the other, that product equals the product of the other numerator multiplied by the other denominator. So to solve* $\frac{2}{x} = \frac{4}{6}$, *you would cross-multiply to get* $2(6) = 4x$, *or* $x = 3$.

Note *The* ***opposite*** *of a term is the same term with its opposite sign. So the opposite of* $3x$ *is* $-3x$, *the opposite of* $-7y$ *is* $+7y$, *and the opposite of 0 is 0. The technical name for* opposite *is* ***additive inverse****.*

Note *By the x term, I mean the term with the unknown value in it that you are trying to find. It could be any other letter, but it's usually an x.*

Note *In the following four examples, the step numbers will refer to the list of steps shown above. For each example, several numbered steps may be omitted because they are not needed to find the solution.*

Example 1: Solve for x: $7x - 2 = 10x + 13$

Solution:

$7x - 2 = 10x + 13$	Step 5: Add $-10x$ to each side.
$-3x - 2 = +13$	Step 6: Add $+2$ to each side.
$-3x = 15$	Step 8: Divide each side by -3.
$x = -5$	Solution.

Example 2: Solve for x: $7 = 2(3x - 5) - 4(x - 6)$

Solution:

$7 = 2(3x - 5) - 4(x - 6)$	Step 2: Switch sides.
$2(3x - 5) - 4(x - 6) = 7$	Step 3: Multiply out the parentheses.
$6x - 10 - 4x + 24 = 7$	Step 4: Combine like terms on each side.
$2x + 14 = 7$	Step 6: Add -14 to each side.
$2x = -7$	Step 8: Divide each side by 2.
$x = -\frac{7}{2}$	Solution. The answer doesn't have to be an integer.

Example 3: Solve for x: $\frac{x}{4} + \frac{x}{6} = 1$

Solution:

$\frac{x}{4} + \frac{x}{6} = 1$	Step 1: Multiply each term by 12.
$3x + 2x = 12$	Step 4: Combine like terms.
$5x = 12$	Step 8: Divide each side by 5.
$x = \frac{12}{5}$	Solution.

Example 4: Solve for x: $\frac{4}{x-7} = \frac{5}{x+2}$.

Solution:

$\frac{4}{x-7} = \frac{5}{x+2}$	Step 1: Cross-multiply.
$4(x+2) = 5(x-7)$	Step 3: Distribute.
$4x + 8 = 5x - 35$	Step 5: Add $-5x$ to each side.
$-x + 8 = -35$	Step 6: Add -8 to each side.
$-x = -43$	Step 8: Divide each side by -1.
$x = 43$	Solution

Note *The answer to this problem could not be 7 or −2. Each would make one of the original fractions equal zero on the bottom. If one of these two numbers were the only possible answer, then the problem would have no answer because you are not allowed to divide by zero.*

Example 5: Solve for x: $y = \frac{3x-5}{x-7}$

Solution:

Write $y = \frac{y}{1}$: $\frac{y}{1} = \frac{3x-5}{x-7}$	Step 1: Cross-multiply.
$(x-7)y = 1(3x-5)$	Step 3: Distribute.
$xy - 7y = 3x - 5$	Step 5: Add $-3x$ to each side.
$xy - 3x - 7y = -5$	Step 6: Add $7y$ to each side.
$xy - 3x = 7y - 5$	Step 7: Factor out the x from the left.
$x(y-3) = 7y - 5$	Step 8: Divide each side by $y - 3$.
$x = \frac{7y-5}{y-3}$	Solution.

Example 6: Solve for x:

Problem	Solution
a. $4x - 9 = 2x - 6$	$\frac{3}{2}$
b. $4x + 5 + 6 + 7x = 0$	-1
c. $4(5x-6) = 7(8x-9)$	$\frac{13}{12}$
d. $10(2x-3) - 30x = 45$	-7.5
e. $ax + b = c$	$\frac{c-b}{a}$

f. $\frac{1}{x} + \frac{1}{5} = \frac{1}{2}$ $\qquad \frac{10}{3}$

g. $y = \frac{3}{x - 5}$ $\qquad \frac{3 + 5y}{y}$

h. $(x + 5)(y + z) = 6$ $\qquad \frac{6 - 5y - 5z}{y + z}$

i. $(x + 2)(x + 3) - x(x + 8) = 24$ $\qquad -6$

j. $7x = 10x$ $\qquad 0$

Q Let's do some multiple-choice exercises.

Exercise 1: $2x - 6 = 4; x + 3 =$

A. 5
B. 6
C. 7
D. 8
E. 9

Exercise 2: $x - 9 = 9 - x; x =$

A. 0
B. 4.5
C. 9
D. 13.5
E. 18

Exercise 3: $4x - 17 = 32; 12x - 51 =$

A. $12\frac{1}{4}$
B. $36\frac{3}{4}$
C. 64
D. 96
E. 288

Exercise 4: $\frac{xy}{y - x} = 1; x =$

A. $\frac{1}{2}$
B. 1
C. $\frac{y}{y + 1}$
D. $\frac{y}{y - 1}$
E. $\frac{y}{1 - y}$

Let's look at the answers.

Answer 1: **D:** $x = 5$, but the question asks for $x + 3 = 8$.

Answer 2: **C:** $2x = 18$; $x = 9$. The answer is C. It actually can be solved just by looking because $9 - 9 = 9 - 9$ (or $0 = 0$).

Answer 3: **D:** We do not solve this equation. We recognize that $12x - 51 = 3(4x - 17) = 3(32) = 96$.

Answer 4: **C:** By cross-multiplying, we get $xy = y - x$. Then $xy + x = y$, which factors to $x(y + 1) = y$. Therefore, $x = \frac{y}{y + 1}$.

LINEAR INEQUALITIES

To review some facts about inequalities:

$a < b$ (read, "a is less than b") means a is to the left of b on the number line.

$a > b$ (read, "a is greater than b") means a is to the right of b on the number line.

$x > y$ is the same as $y < x$.

The notation $x \geq y$ (read, "x is greater than or equal to y") means $x > y$ or $x = y$.

Similarly, $x \leq y$ (read, "x is less than or equal to y") means $x < y$ or $x = y$.

We solve linear inequalities ($<, >, \leq, \geq$) the same way we solve linear equalities, except when we multiply or divide by a negative, the inequality sign reverses direction.

Example 7: Solve for x: $6x + 2 < 3x + 10$

Solution: $3x < 8$, so $x < \frac{8}{3}$. The inequality does not switch because both sides are divided by a positive number (3).

Example 8: Solve for x: $-2(x - 3) \leq 4x - 3 - 7$

Solution: $-2x + 6 \leq 4x - 10$, or $-6x \leq -16$. Thus, $x \geq \frac{-16}{-6} = \frac{8}{3}$. Here the inequality switches because we divided both sides by a negative number (-6).

Example 9: Solve for x: $8 > \frac{x-2}{-3} \geq 5$

Solution: We multiply through by -3, and both inequalities switch. We get $-24 < x - 2 \leq -15$. If we add 2 to each part to get a value for x alone, the final answer is $-22 < x \leq -13$.

Note *This answer is read as "x is greater than −22 and less than or equal to −13." This type of "double" inequality occurs when x is between two values.*

Note *Sometimes the x value can be outside of two values, such as $x < -3$ or $x > 2$, read as "x is less than −3 or greater than 2." Use "or" because x cannot be both at the same time.*

Example 10: Solve for x:

Problem	Solution
a. $10x - 5 < 5x - 11$	$x < -\frac{6}{5}$
b. $-2x - 7 < 4x + 9$	$x > -\frac{8}{3}$
c. $4(2x - 3) - 7x \leq 12 - 2x$	$x \leq 8$
d. $1 - 2x - 7 \leq -6$	$x \geq 0$
e. $1 \leq 2(3x - 5) < 7$	$\frac{11}{6} \leq x < \frac{17}{6}$
f. $1 < \frac{2x-5}{-3} \leq 7$	$1 > x \geq -8$ *or* $-8 \leq x < 1$

Q **Now, let's do some more exercises.**

Exercise 5: If $3x + 4 > 17$, $3x + 7 >$

A. $\frac{13}{3}$

B. $\frac{22}{3}$

C. 14

D. 17

E. 20

Exercise 6: If $3x + 4y < 5$, $x <$

A. $5 - 4y - 3$

B. $\frac{5}{4}y - 3$

C. $\frac{5}{12}y$

D. $\frac{5-4y}{3}$

E. $\frac{5}{3} - 4y$

Exercise 7: $x > 0$ and $y > 0$. The number of ordered pairs of whole numbers (x, y) such that $2x + 3y < 9$ is

A. 1
B. 2
C. 3
D. 4
E. 5

Let's look at the answers.

Answer 5: **E:** We don't actually have to solve this one. $3x + 7 = (3x + 4) + 3 > 17 + 3$, or 20.

Answer 6: **D:** $3x + 4y < 5$ is the same as $3x < 5 - 4y$. Dividing by 3, we get $\frac{5 - 4y}{3}$.

Answer 7: **C:** We must substitute numbers. (1, 1) is okay because $2(1) + 3(1) < 9$; (2, 1) is okay because $2(2) + 3(1) < 9$; (1, 2) is okay because $2(1) + 3(2) < 9$; and that's all.

We will do more on ordered pairs later in the book. As we have seen already, some questions overlap more than one topic.

ABSOLUTE VALUE EQUALITIES

Absolute value is the magnitude of a number, without regard to sign. You should know the following facts about absolute value:

$|3| = 3$, $|-7| = 7$, and $|0| = 0$

$|u| = 6$ means that $u = 6$ or -6.

$|u| = 0$ always means $u = 0$.

$|u| = -17$ has no solutions because the absolute value is never negative.

Example 11: Solve for x: $|2x - 5| = 7$

Solution: Either $2x - 5 = 7$ or $2x - 5 = -7$. So $x = 6$ or $x = -1$.

Note *This kind of problem always has two answers.*

Example 12: Solve for x: $|5x + 11| = 0$.

Solution: $5x + 11 = 0$; $x = -\frac{11}{5}$.

Note *This type of problem (absolute value equals 0) always has one answer.*

Those of you with some math background know there is a lot more to absolute value. This section and the next, however, are all you need for the ACCUPLACER.

ABSOLUTE VALUE INEQUALITIES

If we talk about integers $|u| \leq 3$, we have $|u| = -3, -2, -1, 0, 1, 2$, and 3. So $|u| \leq a$ means $-a \leq u \leq a$, where $a > 0$.

Example 13: Solve for x: $|2x - 3| < 11$

Solution: $-11 < 2x - 3 < 11$. Adding 3 to each piece means $-8 < 2x < 14$. Dividing by 2 gives $-4 < x < 7$.

If we talk about integers $|u| \geq 4$, we have $u = 4, 5, 6, \ldots$ and $-4, -5, -6, \ldots$. So $|u| \geq a$ means $u \geq a$ or $u \leq -a$, $a > 0$.

Example 14: Solve for x: $|x - 7| > 4$.

Solution: $x - 7 > 4$ or $x - 7 < -4$. The two parts of the answer are $x > 11$ or $x < 3$.

Example 15: Solve for x:

Problem	Solution
a. $\|3x - 5\| = 7$	$x = 4, -\frac{2}{3}$
b. $\|5 - 3x\| = 7$	$x = 4, -\frac{2}{3}$
c. $\|4x + 7\| = 3$	$x = -1, -\frac{5}{2}$
d. $\|7x + 3\| = 0$	$x = -\frac{3}{7}$
e. $\|7x - 2\| = -2$	No answer; the absolute value cannot be negative.
f. $\|4x - 7\| < 13$	$-\frac{3}{2} < x < 5$
g. $\|3x + 5\| > 10$	$x > \frac{5}{3}$ or $x < -5$
h. $\|2x - 11\| > -7$	All real numbers; the absolute value is always larger than any negative.
i. $\|5x + \pi\| < -2$	No answer; the absolute value cannot be negative.
j. $3 \leq \|2x - 7\| \leq 3$	$x = 2, 5$

Let's do some more multiple-choice exercises:

Exercise 8: $|2x + 1| = |x + 5|$; $x =$

A. -2
B. 0
C. 4
D. 4 and -2
E. 0 and 4

Exercise 9: $|x - y| = |y - x|$. This statement is true

A. for no values
B. only if $x = y = 0$
C. only if $x = y$
D. only for all integers
E. for all real numbers

Exercise 10: If $4|2x + 3| = 11$, $8|2x + 3| + 5 =$

A. $\frac{11}{4}$
B. $\frac{31}{4}$
C. 22
D. 27
E. 110

Exercise 11: If $5 \leq |x| \leq 5$, $x =$

A. 0
B. 5
C. -5 and 5
D. no values
E. all values

Exercise 12: $|x - 5| \geq -5$ if $x =$

A. 0
B. 5
C. -5 and 5
D. no values
E. all values

Exercise 13: $|x + 5| \leq -5$ if $x =$

A. 0
B. 5
C. -5 and 5
D. no values
E. all values

Let's look at the answers.

Answer 8: **D:** $2x + 1 = x + 5$ or $2x + 1 = -(x + 5)$, for which the answers are 4 and -2, respectively.

Answer 9: **E:** For example, if $x = 3$ and $y = -2$, $|3 - (-2)| = |-2-3| = 5$.

Answer 10: **D:** We don't have to solve this for x at all. If $4|2x + 3| = 11$, then $8|2x + 3| = 2(11) = 22$. Adding 5, we get 27.

Answer 11: **C:** Both $|-5|$ and $|5|$ equal 5.

Answer 12: **E:** The absolute value is always greater than any negative number because it is always greater than or equal to zero.

Answer 13: **D:** The absolute value cannot be less than a negative number.

QUADRATIC EQUATIONS

Quadratic equations, equations involving the square of the variable, can be solved in three principal ways: factoring, taking the square root, and using the quadratic formula. Another name for the solution of any equation is a **root**. An equation has as many roots as the highest power of the unknown variable.

Solving Quadratics by Factoring

Solving quadratic equations by factoring is based on the fact that if $a \times b = 0$, then either $a = 0$ or $b = 0$.

Example 16: Solve for all values of x: $x(x - 3)(x + 7)(2x + 1)(3x - 5)(ax + b)(cx - d) = 0$.

Solution: Setting each factor equal to 0 (better if you can do it just by looking), we get $x = 0, 3, -7, -\frac{1}{2}, \frac{5}{3}, -\frac{b}{a}$, and $\frac{d}{c}$.

Solving Quadratics by Taking the Square Root

If the equation is of the form $x^2 = c$, with no x term, we just take the square root: $x = \pm\sqrt{c}$.

Note *Remember that $\sqrt{9} = 3$, $-\sqrt{9} = -3$, $\sqrt{-9}$ is not real, and if $x^2 = 9$, then $x = \pm 3$!*

Example 17: Solve for all values of x:

a. $x^2 - 7 = 0$

b. $ax^2 - b = c$, where $a, b, c > 0$.

Solutions: **a.** $x^2 = 7$; $x = \pm\sqrt{7}$

b. $ax^2 = b + c$; $x^2 = \dfrac{b+c}{a}$, so $x = \pm\sqrt{\dfrac{b+c}{a}}$, or $\dfrac{\pm\sqrt{a(b+c)}}{a}$

Example 18: Solve for x by factoring:

Problem	Solution
a. $x^2 - 5x - 6 = 0$	$x = 6, -1$
b. $x^2 - 3x - 40 = 0$	$x = 8, -5$
c. $3x^2 - 10x + 3 = 0$	$x = \frac{1}{3}, 3$
d. $x^3 + 7x^2 - 8x = 0$	$x = 1, 0, -8$
e. $x^2 - 7x + 9 = 3$	$x = 1, 6$
f. $2x^3 - 18x = 0$	$x = 3, 0, -3$
g. $x^4 - 10x^2 + 9 = 0$	$x = \pm 1, \pm 3$
h. $x^4 - 8x^2 - 9 = 0$	$x = \pm 3$

Example 19: Solve for x by taking the square root:

Problem	Solution
a. $x^2 = 9$	$x = \pm\sqrt{9} = \pm 3$
b. $x^2 = 11$	$x = \pm\sqrt{11}$
c. $5x^2 - 16 = 0$	$x = \dfrac{\pm 4\sqrt{5}}{5}$
d. $Ax^2 + M + I = T$	$x = \pm\dfrac{\sqrt{A(T - I - M)}}{A}$
e. $(mx + n)^2 = p$	$x = \dfrac{\pm\sqrt{p} - n}{m}$
f. $x^2 = -100$	$x = \pm 10i$
g. $5x^2 + 7 = 0$	$x = \dfrac{\pm\sqrt{35}}{5}i$

Q Let's do some more multiple-choice exercises:

Exercise 14: The solutions to $x^2 - 3x - 4 = -6$ are

A. $x = 4$ and $x = -1$
B. $x = 4$ and $x = 1$
C. $x = 1$ and $x = -1$
D. $x = 1$ and $x = 2$
E. $x = 5$ and $x = -2$

Exercise 15: If $x = 6$ and $x = -9$ are the answers to a quadratic equation, the original quadratic equation is

A. $x^2 + 15x - 54 = 0$
B. $x^2 - 15x - 54 = 0$
C. $x^2 - 3x - 54 = 0$
D. $x^2 + 3x - 54 = 0$
E. $x^2 - 54x + 54 = 0$

Exercise 16: $x = 2a$ and $x = -5b$ are solutions to a quadratic equation. The quadratic equation is

A. $x^2 - 2ab - 10ab = 0$
B. $x^2 + 2ab - 10ab = 0$
C. $x^2 + (2a - 5b)x - 10ab = 0$
D. $x^2 - (2a - 5b)x - 10ab = 0$
E. $x^2 + (2x - 5b)x + 10ab = 0$

A Let's look at the answers.

Answer 14: **D:** Add 6 to both sides and factor to get $x^2 - 3x + 2 = (x - 2)(x - 1) = 0$. Then set each factor equal to 0.

Answer 15: **D:** If 6 is a solution, $(x - 6)$ is a factor, and if -9 is a solution, $(x + 9)$ is a factor. So $(x - 6)(x + 9) = x^2 + 3x - 54 = 0$.

Answer 16: **D:** $(x - 2a)(x + 5b) = x^2 - 2ax + 5bx - 10ab = x^2 - (2a - 5b)x - 10ab = 0$.

Solving Quadratics by Using the Quadratic Formula

The quadratic formula states that if $ax^2 + bx + c = 0$, then

$$x = \frac{-b \pm \sqrt{b^2 - 4ac}}{2a},$$

where a is the coefficient of the x^2 term, b is the coefficient of the x term, and c is the number term.

Example 20: Solve $3x^2 - 5x + 2 = 0$ by using the quadratic formula.

Solution: $a = 3, b = -5, c = 2$. $x = \dfrac{-(-5) \pm \sqrt{(-5)^2 - 4(3)(2)}}{2(3)} = \dfrac{5 \pm 1}{6}$. x_1 (read as "x sub one," the first answer) $= \dfrac{5+1}{6} = 1$; x_2 (read, "x sub two," the second answer) $= \dfrac{5-1}{6} = \dfrac{2}{3}$.

Example 21: Solve $3x^2 - 5x + 2 = 0$ by factoring.

Solution: This is the same problem as Example 20. $3x^2 - 5x + 2 = (3x - 2)(x - 1) = 0$, so $x = \dfrac{2}{3}$ or 1.

Factoring is preferred; using the quadratic formula takes too long.

Before we go to the next set of exercises, I suppose most of you know the quadratic formula, but few have seen it shown to be true. The teacher in me has to show you, even if you don't care.

$ax^2 + bx + c = 0$ — The coefficient of x^2 must be 1.

$x^2 + \dfrac{b}{a}x = -\dfrac{c}{a}$ — Complete the square; this means taking half the coefficient of x, squaring it, and adding it to both sides.

$x^2 + \dfrac{b}{a}x + \left(\dfrac{b}{2a}\right)^2 = \left(\dfrac{b}{2a}\right)^2 - \dfrac{c}{a}$ — Factor the left side and do the algebra on the right side.

$\left(x + \dfrac{b}{2a}\right)^2 = \dfrac{b^2}{4a^2} - \dfrac{c}{a}\dfrac{(4a)}{(4a)} = \dfrac{b^2 - 4ac}{4a^2}$ — Multiply the last term by $\dfrac{4a}{4a}$ and combine terms.

$\left(x + \dfrac{b}{2a}\right)^2 = \dfrac{b^2}{4a^2} - \dfrac{4ac}{4a^2} = \dfrac{b^2 - 4ac}{4a^2}$ — Take the square root of both sides.

$x + \dfrac{b}{2a} = \dfrac{\pm\sqrt{b^2 - 4ac}}{2a}$ — Solve for x and simplify.

$x = -\dfrac{b}{2a}\dfrac{\pm\sqrt{b^2 - 4ac}}{2a} = \dfrac{-b \pm \sqrt{b^2 - 4ac}}{2a}$

The quadratic formula is really true!

We now know for sure that if $ax^2 + bx + c = 0$, then $x = \dfrac{-b \pm \sqrt{b^2 - 4ac}}{2a}$, where a is the coefficient of x^2, b is the coefficient of x, and c is the non-x term. Let's make sure you really understand this.

Example 22: For each problem, tell what a is, what b is, and what c is.

Problem	Solution		
	a	b	c
a. $3x^2 - 4x - 7 = 0$	3	-4	-7
b. $x^2 - 15 = 0$	1	0	-15
c. $x^2 + 2x + \frac{7}{4} = 0$	1	2	$\frac{7}{4}$
d. $3x - x^2 = 0$	-1	3	0
	(Rewrite as $-x^2 + 3x = 0$.)		
e. $a^2x^2 - bx - c = 0$	a^2	$-b$	$-c$
	(The coefficient of $x^2 = a^2$; in front of x is $-b - c$, (not its coefficient,) is the term without x!)		
f. $-x^2 + b^2 + 7 = 0$	-1	0	$b^2 + 7$
	(No x term; two terms without x.)		
g. $3x^2 + ax + 4x - 2 = 0$	3	$a + 4$	-2
	(Rewrite $ax + 4x$ as $(a + 4)x$; coefficient of x is $(a + 4)$.)		
h. $2x^2 - ix + 3 = 0$	2	$-i$	3
i. $3ix^2 - 2x - 5i = 0$	$3i$	-2	$-5i$
j. $y = x^2 - 2x + 3$	1	-2	$3 - y$
	(Rewrite as $x^2 - 2x + (3 - y) = 0$.)		

Example 23: For each of the problems in Example 22, solve for x by using the quadratic formula.

Problem	Solution
a. $3x^2 - 4x - 7 = 0$	$x = \dfrac{-(-4) \pm \sqrt{(-4)^2 - 4(3)(-7)}}{2(3)} = -1, \dfrac{7}{3}$
b. $x^2 - 15 = 0$	$x = \dfrac{0 \pm \sqrt{0^2 - 4(1)(-15)}}{2(1)} = \pm\sqrt{15}$
c. $x^2 + 2x + \frac{7}{4} = 0$	$x = \dfrac{-2 \pm \sqrt{2^2 - 4(1)\left(\frac{7}{4}\right)}}{2(1)} = \dfrac{-2 \pm i\sqrt{3}}{2}$
	(Answer is two imaginary numbers.)

d. $3x - x^2 = 0$

$$x = \frac{-3 \pm \sqrt{3^2 - 4(-1)(0)}}{2(-1)} = 0, 3$$

(On a test, you would solve this one by factoring.)

e. $a^2x^2 - bx - c = 0$

$$x = \frac{-(-b) \pm \sqrt{(-b)^2 - 4(a^2)(-c)}}{2a^2}$$

$$= \frac{b \pm \sqrt{b^2 + 4a^2c}}{2a^2}$$

(Ugly, but correct.)

f. $-x^2 + b^2 + 7 = 0$

$$x = \frac{0 \pm \sqrt{0^2 - 4(-1)(b^2 + 7)}}{2(-1)} = \pm\sqrt{b^2 + 7}$$

(Ugly again; the higher you go in math, the uglier the answers can get (not always, but more often!))

g. $3x^2 + ax + 4x - 2 = 0$

$$\frac{-(a + 4) \pm \sqrt{(a + 4)^2 - 4(3)(-2)}}{2(3)}$$

$$= \frac{-a - 4 \pm \sqrt{a^2 + 8a + 40}}{6}$$

h. $2x^2 - ix + 3 = 0$

$$x = \frac{-(-i) \pm \sqrt{(-i)^2 - 4(2)(3)}}{2(2)}$$

$$= \frac{i \pm 5i}{4} = \frac{3}{2}i, -i$$

i. $3ix^2 - 2x - 5i = 0$

$$x = \frac{-(-2) \pm \sqrt{(-2)^2 - 4(3i)(-5i)}}{2(3i)}$$

$$= \frac{2 \pm \sqrt{-56}}{6i} = \frac{2 \pm 2i\sqrt{14}}{6i} \times \frac{-i}{-i}$$

$$= \frac{\pm\sqrt{14} - i}{3} = \frac{\pm\sqrt{14}}{3} - \frac{1}{3}i$$

j. $y = x^2 - 2x + 3$

$$x = \frac{-(-2) \pm \sqrt{(-2)^2 - 4(1)(3 - y)}}{2(1)}$$

$$= \frac{2 \pm \sqrt{4y - 8}}{2} = 1 \pm \sqrt{y - 2}$$

(This last problem is used to find the inverse map with functions. The values of x determine whether we use the + or − sign. We will have more about this later.)

QUADRATIC INEQUALITIES

This particular section is one of the few topics that is taught much better now than when I learned it in high school several hundred years ago. We will do it here in a way that also may be better than the way you were taught.

Example 24: Solve for x: $x^2 - 2x - 80 > 0$.

Solution: In solving any inequalities, keep in mind that the answers are in regions. The first step we do here is to move all of the terms to the left side, leaving 0 on the right, and factor the left side of the inequality.

For this inequality, we get $(x - 10)(x + 8) > 0$. If we set each factor equal to 0, we get $x = 10$ and $x = -8$, which form the boundaries of the regions. There are three regions here: $x > 10$; $-8 < x < 10$; and $x < -8$, as shown in the figure. We then substitute a number from each region into the original inequality. (You check only one number in each region.) We are not looking for the answer here; we are just looking for the *sign* of the answer.

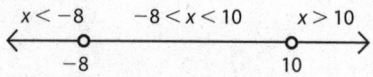

Note *The open circles at −8 and 10 mean they cannot be part of the answer, since the product ≠ 0. Some books use parentheses instead of open dots.*

1. For $x > 10$; let's use $x = 20$: $(20 - 10)$ is positive; $(20 + 8)$ is positive; the product of two positives is positive, which is greater than 0. So the region $x > 10$ is part of the answer.
2. For $-8 < x < 10$, let's use $x = 3$: $(3 - 10)$ is negative; $(3 + 8)$ is positive; the product of a negative and positive is *not* greater than 0; so $-8 < x < 10$ is not part of the answer.
3. For $x < -8$, let's use $x = -100$ (it doesn't have to be a close number, just in the region). $(-100 - 10)$ is negative; $(-100 + 8)$ is negative; the product of two negatives is positive, which is greater than 0. So $x < -8$ is part of the answer.

So the answer is $x > 10$ or $x < -8$.

Graphing this answer on the number line, we get the following:

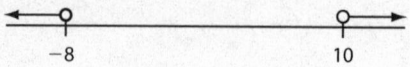

Example 25: Solve for x: $x^2 + 5x \leq 14$.

Solution: We must get everything to one side with the coefficient of the highest term positive to make factoring easier. We get $x^2 + 5x - 14 \leq 0$, or $(x + 7)(x - 2) \leq 0$. Setting each factor equal to 0, we see that $x = -7$ or $x = 2$. The intervals are $x \geq 2$, $-7 \leq x \leq 2$, and $x \leq -7$. Let's take the numbers 5, 0, and -10 in the three intervals. First, $(5 + 7)$ and $(5 - 2)$ are positive; their product is positive, which is not less than or equal to 0, so the region $x \geq 2$ is *not* part of the answer. Next, $(0 + 7)$ is positive and $(0 - 2)$ is negative; their product is negative, so $-7 \leq x \leq 2$ is part of the answer. Finally, $(-10 + 7)(-10 - 2)$ is the product of two negatives, which is positive and not less than or equal to 0; so $x \leq -7$ is not part of the answer. So the answer is $-7 \leq x \leq 2$, and its graph looks like this:

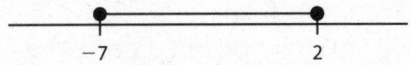

Note *The solid dots indicate that -7 and 2 are part of the answer. In some books, this inclusion is indicated by brackets.*

Example 26: Solve for x: $\dfrac{x-2}{x-4} \leq \dfrac{x-8}{x-6}$.

Before we start this problem, let's check a couple of notes.

Note *This is a rather long problem. If you did this problem like I did in high school, it would take nearly forever.*

Note *If this were an equation, you would cross-multiply to solve. In this problem cross-multiplication is wrong, wrong, wrong! When you multiply by an x on the bottom, the inequality sign may or may not switch, depending on the value for x.*

Solution: We start by getting everything to one side by first subtracting the right side from both sides and simplifying.

$$\frac{x-2}{x-4} - \frac{x-8}{x-6} = \frac{(x-2)(x-6)-(x-4)(x-8)}{(x-4)(x-6)} = \frac{4(x-5)}{(x-4)(x-6)} \leq 0$$

So $x - 4 = 0$ means $x = 4$; $x - 5 = 0$ means $x = 5$, and $x - 6 = 0$ means $x = 6$. We therefore check the four intervals, $x < 4$, $4 < x \leq 5$, $5 \leq x \leq 6$, and $x > 6$. Let's take the numbers 2, 4.5,

5.5, and 10. Determining the sign of each of the intervals (remember you *never* do the arithmetic, *NEVER*!), we see that $x < 4$ and $5 \le x \le 6$ is the answer. Its graph looks like this:

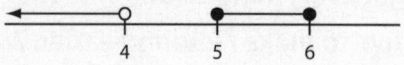

Note *A fraction equals 0 when the numerator equals 0 and the denominator does not equal zero. On the graph, the dot on 5 is solid and the dots on 4 and 6 are open.*

Example 27. Solve for x and graph the solution:

Problem	Solution	Graph of the Solution
a. $x^2 - 8x + 7 > 0$	$x > 7$ or $x < 1$	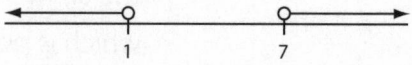
b. $x^2 + 2x < 80$	$-10 < x < 8$	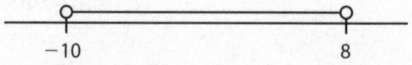
c. $2x^2 + 4x - 6 \ge 0$	$x \le -3$ or $x \ge 1$	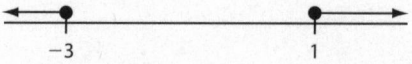
d. $16 - x^2 \le 0$	$x \ge 4$ or $x \le -4$	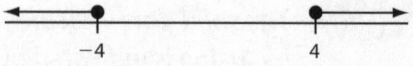
e. $x^3 - 6x^2 + 8x > 0$	$x > 4$ or $0 < x < 2$	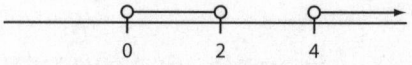
f. $2x^3 + 12x^2 \le -10x$	$x \le -5$ or $-1 \le x \le 0$	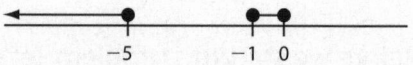
g. $\dfrac{3x-5}{x-2} \ge 2$	$x \le 1$ or $x > 2$	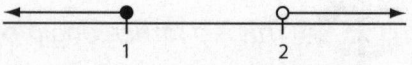
h. $\dfrac{x-5}{x-3} - \dfrac{x-7}{x-1} \le 0$	$x < 1$ or $3 < x \le 4$	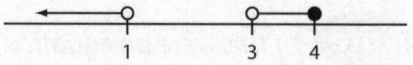

Now, let's do the chapter review test.

Chapter 6 Test

In each question, solve for x.

1. $3x - 9 = -7x + 33$.
2. $4(2x - 3) - 5(4x - 1) = 6$.
3. $ax + by = c$.
4. $y = \dfrac{3x - 5}{x - 2}$.

5. $\frac{4}{x-7} = \frac{7}{x-4}$.
6. $7x^2 - 19 = 0$.
7. $5x^2 - x + 10 = 0$.
8. $|3x - 7| = 5$.
9. $4x - 6 < 7x + 24$.
10. $|2x - 4| > 10$.
11. $x^4 - 26x^2 + 25 = 0$.
12. $x^2 - 5x > 36$.
13. $\frac{5x-14}{x-2} \leq 3$.
14. $\frac{x}{x-2} \geq \frac{x}{x+2}$
15. $cx^2 - bx + a = 0$.
16. $|2x - 5| = |4x - 1|$.
17. $-4 < -2x - 6 \leq 10$.
18. $m(x - y) + r = s$.
19. $x^3 - 20x^2 = 21x$.
20. $9x = 11x$.
21. $x^2 - 8x > -16$
22. $x^2 + 20x + 100 \leq 0$

Answers to Chapter 6 Test

1. 4.2
2. $-\frac{13}{12}$.
3. $\frac{c-by}{a}$.
4. $\frac{2y-5}{y-3}$.
5. 11.
6. $\frac{\pm\sqrt{133}}{7}$.

7. $\frac{1 \pm i\sqrt{199}}{10}$.

8. $4, \frac{2}{3}$.

9. $x > -10$.

10. $x > 7$ or $x < -3$.

11. $\pm 1, \pm 5$.

12. $x > 9$ or $x < -4$.

13. $2 < x \leq 4$.

14. $x > 2$ or $-2 < x \leq 0$.

15. $\frac{b \pm \sqrt{b^2 - 4ac}}{2c}$.

16. $-2, 1$.

17. $-8 \leq x < -1$.

18. $\frac{s - r + my}{m}$.

19. $-1, 0, 21$.

20. 0.

21. All real numbers: $x^2 - 8x + 16 = (x - 8)^2 > 0$ is never negative and can't be 0 here.

22. Only $x = -10$ since $(x + 10)^2$ is never negative but can $= 0$.

Let's go on.

Answer to "Bob Asks": Asphyxiate.

CHAPTER 7: ***Words and Word Problems***

"What word contains three double letters in a row?"

In a certain way, this was the hardest chapter to write. This area is very necessary and is severely neglected in high school. The placement test will not ask very many word problems other than the ones we have seen for arithmetic. Yet if you go on in math, not seeing these problems will be very costly to you. I decided to include here the most likely ones you might see. If you want more, go to my *Algebra* book for the ones you didn't see, my *Precalculus* book for a preview for calculus, my *Calc 1* and *Calc 3* books for those type of word problems, my *GRE* book when you go to a graduate or business school, and my *GMAT* book when going for your MBA. These books are listed in the beginning of this book.

Too much talking! Let's get started.

BASICS

As we know, the answer in **addition** is the **sum**. Other words that indicate addition are **plus**, **more**, **more than**, **increase**, and **increased by**. You can write all sums in any order because addition is commutative.

The answer in **multiplication** is the **product**. Another word that is used is **times**. Sometimes the word **of** indicates multiplication, as we shall see shortly. **Double** means to multiply by two, and **triple** means to multiply by three. Because multiplication is also commutative, we can write any product in any order.

Division's answer is called the **quotient**. Another phrase that is used is **divided by**.

The answer in **subtraction** is called the **difference**. Subtraction can present a reading problem because $4 - 6 \neq 6 - 4$, so we must be careful to subtract in the correct order. Example 1 shows how some subtraction phrases are translated into algebraic expressions.

Example 1:

Phrases	Expressions
a. The difference between 9 and 5	$9 - 5$
The difference between m and n	$m - n$
b. Five minus two	$5 - 2$
m minus n	$m - n$
c. Seven decreased by three	$7 - 3$
m decreased by n	$m - n$
d. Nine diminished by four	$9 - 4$
m diminished by n	$m - n$
e. Ten less two	$10 - 2$
m less n	$m - n$
f. Ten less than two	$2 - 10$
m less than n	$n - m$
g. Three from five	$5 - 3$
m from n	$n - m$

Note *One word can make a big difference. In parts e and f of Example 1, m less n means $m - n$; and m less than n means $n - m$. In addition, m is less than n means $m < n$. You must read carefully!*

The following words usually indicate an equal sign: *is, am, are, was, were, the same as, equal to.*

You also must know the following phrases for inequalities: at least ($\geq$), not more than ($\leq$), over ($>$), and under ($<$).

Example 2: Write the following in symbols:

Problem	Solution
a. m times the sum of q and r	$m(q + r)$
b. Six less than the product of x and y	$xy - 6$
c. The difference between c and d divided by f	$\frac{c - d}{f}$
d. b less than the quotient of r divided by s	$\frac{r}{s} - b$
e. The sum of d and g is the same as the product of h and r	$d + g = hr$
f. x is at least y	$x \geq y$
g. Zeb's age n is not more than 21	$n \leq 21$
h. I am over 30 years old, where "I" represents "my age"	$I > 30$
i. Most people are under seven feet tall, where "p" represents "most people"	$p < 7$

COMMON AND EVERYDAY PROBLEMS

Example 3: In spring at a recent school board election, Ms. Johnson received 50 more votes than Ms. Smith for president of the school board. If 892 votes were cast, how many votes did each get?

Solution: In most problems, it is better to let x = the smaller number—in this case, Ms. Smith's vote total. Ms. Johnson received $x + 50$ votes. Then $x + (x + 50) = 892$; $2x = 842$; so $x = 421$, the number of Ms. Smith's votes. Ms. Johnson received $421 + 50 = 471$ votes.

Example 4: Two numbers total 40. The sum of twice one of them and triple the other is 105. Find the numbers.

Note *If two numbers total 40 and one is 11, the other is 40 − 11 (or 29). If two numbers total 40 and one is unknown, x, the other is 40 − x.*

Solution: Let one number $= x$; the other number is $40 - x$. The sum (add) of twice one number, $2x$, and triple the other number, $3(40 - x)$, is 105. The equation is $2x + 3(40 - x) = 105$; $2x + 120 - 3x = 105$; $x = 15$ and, because both numbers were asked for, $40 - x = 40 - 15 = 25$ is the other number.

Example 5: The sum of two numbers is 20. Five times the larger less the smaller is 70. Find the larger number.

Solution: Let $x =$ the number you are looking for—in this case, the larger number. Then $20 - x$ is the smaller number. *Less* means "subtract," so don't change the order. The equation is $5x - (20 - x) = 70$; $6x - 20 = 70$; $x = 15$.

RATIOS

The comparison of two numbers is called a **ratio**. The ratio of 3 to 5 is written two ways: $\frac{3}{5}$ or 3 : 5 (read, "the ratio of 3 to 5").

Example 6: Find the ratio of 5 ounces to 2 pounds.

Solution: The ratio is $\frac{5}{32}$, because 16 ounces are in a pound, and a ratio should have the same measurements.

Example 7: A board is cut into two pieces that are in the ratio of 3 to 4. If the board is 56 inches long, how long is the longer piece?

Solution: If the pieces are in the ratio 3 : 4, we let one piece equal $3x$ and the other $4x$. The equation, then, is $3x + 4x = 56$; so $x = 8$; and the longer piece is $4x = 32$ inches.

I have asked this problem many, many times. Almost no one has ever gotten it correct—not because it is difficult, but because no one does problems like this anymore.

DISTANCE AND SPEED

We are familiar with speed being given in miles per hour (mph), so it is easy to remember that speed $= \frac{\text{distance}}{\text{time}}$, or $r = \frac{d}{t}$, where r stands for rate (the speed). Use this relationship, or the equivalent ones: $d = rt$ or $t = \frac{d}{r}$, to do word problems involving speed.

Example 8: Sue drives for 2 hours at 60 mph and 3 hours at 70 mph. What is her average speed?

Solution: Sue's average speed for the whole trip is given by $r = \frac{d}{t}$, where d is the total distance and t is the total time. Use $d = rt$ for each part of her trip to get the total distance. The total distance is $60(2) + 70(3) = 330$ miles. The total time is 5 hours. So Sue's average speed is $r = \frac{330}{5}$ mph $= 66$ mph.

Note *The average speed is not the simple mean of two or more speeds. It is the total distance divided by the total time.*

Example 9: Don goes 40 mph in one direction and returns at 60 mph. What is his average speed?

Solution: Notice that the problem doesn't tell the distance. It doesn't have to; the distance in each direction is the same, because it is a round trip. We can take any distance, so let's choose 120 miles, the LCM of 40 and 60. Then the time going is $\frac{120}{40} = 3$ hours, and the time returning is $\frac{120}{60} = 2$ hours. The average speed is the total distance divided by the total time, $\frac{2(120)}{3+2} = \frac{240}{5} = 48$ mph.

We actually don't have to choose a number for the distance, however. We could use x. Just for learning's sake, we will do this same problem (Example 9) using x as the distance. The time going is $\frac{x}{40}$, and the time returning is $\frac{x}{60}$. Then we have:

$$\text{Total speed} = \frac{\text{total distance}}{\text{total time}} = \frac{2x}{\frac{x}{40} + \frac{x}{60}} = \frac{120(2x)}{120(\frac{x}{40} + \frac{x}{60})} = \frac{240x}{5x} = 48 \text{ mph}$$

Example 10: A plane leaves Indianapolis traveling west. At the same time, a plane traveling 30 mph faster leaves Indianapolis going east. After two hours the planes are 2,000 miles apart. What is the speed of the faster plane?

Solution: A chart and a picture are best for problems like this.

	r	t	d
W	x	2	$2x$
E	$x + 30$	2	$2(x + 30)$

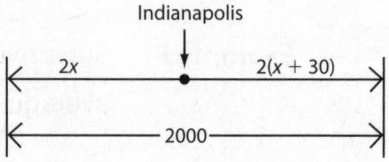

We let x = the speed of the plane going west; then $x + 30$ is the speed of the eastern-going plane. The time for each is 2 hours. Because $rt = d$, the distances are as shown in the above chart. According to the picture, $2x + 2(x + 30) = 2000$. So $x = 485$; $x + 30 = 515$ mph.

The problem is the same if the planes are starting at the ends and flying toward each other.

Q Let's do some basic multiple-choice exercises.

Exercise 1: Four more than a number is seven less than triple the number. The number is

A. 4
B. 5.5
C. 7
D. 9
E. 11

Note *The word "number" does not necessarily mean an integer or even necessarily a positive number.*

Exercise 2: Mike must have at least an 80 average but less than a 90 average to get a *B*. If he received scores of 98, 92, and 75 on the first three tests, which of these grades on the fourth test will give him a *B*?

A. 42
B. 54
C. 66
D. 98
E. 100

Exercise 3: Nine less than a number is the same as the difference between nine and the number. The number is

A. 18
B. 9
C. 0
D. -9
E. All numbers are correct.

Exercise 4: Ed goes 20 mph in one direction and 50 mph on the return trip. His average speed is

A. 25 mph
B. $27\frac{2}{7}$ mph
C. $28\frac{4}{7}$ mph
D. 30 mph
E. $30\frac{6}{7}$ mph

Exercise 5: The angles of a triangle are in the ratio of 3 : 5 : 7. The largest angle is

A. 12°
B. 36°
C. 60°
D. 84°
E. 108°

Exercise 6: The value of d dimes and q quarters in pennies is

A. $d + q$
B. dq
C. $250dq$
D. $10d + 25q$
E. $35dq$

Exercise 7: A car leaves Chicago at 2 . . going west. A second car leaves Chicago at 5 . ., going west at 30 mph faster. At 7 . ., the faster car hits the slower one. The accident occurred after how many miles?

A. 200
B. 100
C. 60
D. 40
E. 20

Exercise 8: A fraction, when reduced, is $\frac{2}{3}$. If 6 is added to the numerator and 14 is added to the denominator, the fraction reduces to $\frac{3}{5}$. The sum of the original numerator and denominator is

A. 60
B. 70
C. 80
D. 90
E. 100

Exercise 9: Fred leaves Fort Worth by car traveling north. Two hours later, Jim also leaves Fort Worth going north, but 20 mph slower. After six more hours, they are 260 miles apart. Fred's speed is

A. 40 mph
B. 50 mph
C. 60 mph
D. 70 mph
E. 80 mph

Exercise 10: The difference in the cost between two books is \$8. Together they cost \$50. The cost of the less expensive book is

A. \$21
B. \$23
C. \$25
D. \$27
E. \$29

Let's look at the answers.

Answer 1: **B:** Let's break this one down into small pieces. Four more than a number is written as $n + 4$ (or $4 + n$). Seven less than triple the number is $3n - 7$ (the only correct way). *Is* means "equals," so the equation is $n + 4 = 3n - 7$. Solving, we get $n = 5.5$.

Answer 2: **C:** The "setup" to do this problem is $80 \leq \frac{(98 + 92 + 75 + x)}{4} < 90$. However, $80(4) = 320$ total points for a minimum, and it must be less than $90(4) = 360$ points. So far, Mike has $98 + 92 + 75 = 265$ points; $265 + 66 = 321$ points. (However, answer choices D or E will result in a grade of A, and I'm sure Mike wouldn't object to that.)

Answer 3: **B:** $x - 9 = 9 - x$; $x = 9$.

Answer 4: **C:** If we assume a 100-mile distance, the original trip was 5 hours, and the return trip was 2 hours. $r = \frac{d}{t} = \frac{200}{7} = 28\frac{4}{7}$mph.

Answer 5: **D:** $3x + 5x + 7x = 180$, so $x = 12°$. The largest angle is $7x = 84°$.

Answer 6: **D:** Dimes are 10 cents each, so the value is $10d$, where d is the number of dimes. Similarly, the value of quarters is $25q$, where q is the number of quarters.

Answer 7: **B:** Let's construct a chart and picture.

	r	t	d
Slower	x	5	$5x$
Faster	$x + 30$	2	$2(x + 30)$

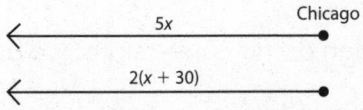

The rate of the slower car is x, and the time of the slower car is 7 . . −2 . ., or 5 hours. The rate of the faster car is $x + 30$; and its time is 7 . . − 5 . ., or 2 hours. When they crashed, their distances were equal, so $5x = 2(x + 30)$. Then $x = 20$; and the total distance is 5(20) or $2(20 + 30) = 100$ miles.

Answer 8: **A:** The fraction can be written as $\frac{2x}{3x}$. So $\frac{2x + 6}{3x + 14} = \frac{3}{5}$. Cross-multiplying, we get $5(2x + 6) = 3(3x + 14)$, so $x = 12$. The original numerator is $2x = 24$ and the original denominator is $3x = 36$, and their sum is $24 + 36 = 60$.

Answer 9:

	r ×	t =	d
Fred	x	8	$8x$
Jim	$x - 20$	6	$6(x - 20)$

D: Fred goes at x mph for 8 hours. Jim goes at $x - 20$ mph for 6 hours. Because they are going in the same direction, we get $8x - 6(x - 20) = 260$, or $x = 70$. Notice that 260 is the *difference* in their distances, not the distance they traveled, so we use subtraction.

Answer 10: **A:** Let x = the less expensive book, so $x + 8$ = the more expensive one; $x + (x + 8) = 50$, so $x = 21$. If you were looking for the more expensive book, you could let x = the more expensive book; then $x - 8$ would be the less expensive one!

MEASUREMENTS

You might want to review some basic measurements and how to convert a few.

Linear: 12 inches = 1 foot; 3 feet = 1 yard; 5,280 feet = 1 mile.

Liquid: 8 ounces = 1 cup; 2 cups = 1 pint; 2 pints = 1 quart; 4 quarts = 1 gallon.

Weight: 16 ounces = 1 pound; 2,000 pounds = 1 ton.

Dry measure: 2 pints = 1 quart; 8 quarts = 1 peck; 4 pecks = 1 bushel. If I love you a bushel and a peck, it would be 5 pecks or 40 dry quarts.

Metric: 1,000 grams in a kilogram; 1,000 milligrams in a gram; 1,000 liters in a kiloliter; 1,000 meters in a kilometer; 1,000 millimeters = 1 meter; 100 centimeters = 1 meter; 10 millimeters = 1 centimeter.

When doing conversions, we pay particular attention to the units, canceling them when doing the multiplication. Sometimes, however, there is confusion about what conversion to use: for example, whether to multiply by $\frac{12 \text{ inches}}{1 \text{ foot}}$ or by $\frac{1 \text{ foot}}{12 \text{ inches}}$. To avoid this confusion, check to see what units are called for in the answer, and then make sure those units are in the correct place in the conversions.

Example 11: Change 30 kilograms 20 grams to milligrams.

Solution: The solution is in milligrams, so be sure milligrams is in the numerator (top) of the conversions. If you need further conversions, such as converting kilograms to grams here, be sure the units will cancel with the conversion for milligrams:

$$\frac{30 \text{ kg}}{1} \times \frac{1000 \text{ g}}{1 \text{ kg}} \times \frac{1000 \text{ mg}}{1 \text{ g}} + \frac{20 \text{ g}}{1} \times \frac{1000 \text{ mg}}{1 \text{ g}} = 30{,}000{,}000 + 20{,}000 = 30{,}020{,}000 \text{ mg}$$

Note *Notice that the measurements g and kg cancel.*

Example 12: Change 90 miles per hour into feet per second.

Solution: The answer is in feet per second, so feet should be in the numerator (top) of any conversion, and seconds should be in the denominator (bottom) of any conversion:

$$\frac{90 \text{ miles}}{\text{hour}} \times \frac{1 \text{ hour}}{60 \text{ minutes}} \times \frac{1 \text{ minute}}{60 \text{ seconds}} \times \frac{5280 \text{ feet}}{1 \text{ mile}} = \frac{132 \text{ feet}}{\text{sec}}$$

Note *Each fraction after the first is equivalent to 1. When we multiply by 1, the value doesn't change. Again, the measurements cancel, and we wind up with feet per second.*

INTEREST

Last, we need to talk a little about simple interest. We know that the interest is equal to the principal times (annual) rate times time (in years). In symbols $i = prt$. The total amount of money is the principal plus the interest, or $A = p + i = p + prt = p(1 + rt)$.

Example 13: Suppose we invest \$20,000 at simple interest at 12% for 3 months. How much money do we have?

Solution: $A = p + prt = 20{,}000 + 20{,}000(.12)\left(\frac{1}{4}\right) = \$20{,}600$. Of course, we wouldn't normally invest at simple interest. However, a safe 12% interest would be great.

Example 14: Suppose we invest \$10,000 at 10% for a year, compounded every six months. How much do we have after a year?

Solution: For the first six months, or half a year, $A = p + prt = 10{,}000 + 10{,}000(.10)\left(\frac{1}{2}\right) = \$10{,}500$. After the second six months, $A = p + prt = 10{,}500 + 10{,}500(.10)\left(\frac{1}{2}\right) = \$11{,}025$, the amount after one year. This is \$25 more than if we had simple interest for the full year. That is why compounding continuously is most desirable—we literally earn interest on our interest.

Chapter 7 Test

For questions 1–12: Write the answer in the form of an equation. Do *not* solve any problems.

1. $2x$ less than y.
2. The product of m and the sum of m and n.
3. The difference of b and c divided by y.
4. The product of v and m from the difference between r and s.
5. Five less than twice a number is the same as 9 less four times the same number.
6. The volume V of a cone is equal to the product of one-third pi, π, the radius r squared, and the height h.
7. The age A of an adult is at least 21.
8. The height h of people in the class is between 66 inches and 80 inches.
9. The area of a trapezoid A equals one-half the product of the height h and the sum of the bases b_1 and b_2.
10. The amount of money A equals the sum of the principal p and the product of the principal, the rate r, and the time t.
11. Let $x =$ the larger number and $y =$ the smaller number. Write the expression that says the sum of twice the larger number and triple the smaller number is the same as the product of the two numbers.
12. Let $x =$ the larger number and $y =$ the smaller number. Write the expression that says the quotient of the larger number divided by the smaller number is at most 12.
13. Change 4 tons 5 pounds 3 ounces to ounces.
14. Change 543 meters to centimeters and kilometers.
15. A board is divided into the ratio of 4:7:9. If the board is 160 inches long, find the length of each piece.
16. You drove 3 hours at 40 miles per hour and 6 hours at 60 mph. Find your average speed.
17. From problem 10, if you invest \$2,000 for 10 years at 8% interest, find your total after 10 years.
18. One number is 10 more than another. The total is 84. Find the numbers.
19. The cost of an item is \$6. The cost of another item is \$8. If I buy one at \$6, what is the largest number of \$8 items I can buy if I have \$100?
20. You want an average between 90 and 95 inclusive. You received 88, 95, and 99 on three tests. What grades can you get on the fourth test so your grade will be between 90 and 95?
21. Mary did $\frac{1}{5}$ of a job and Don did $\frac{1}{3}$ of the job. How much is left to do?
22. Bob did $\frac{1}{6}$ of the job, May did $\frac{1}{3}$, and Sam did $\frac{1}{2}$ of the job. How much is left to do?

Answers to Chapter 7 Test

1. $y - 2x$.
2. $m(m + n)$.
3. $\frac{b - c}{y}$.
4. $(r - s) - vm$.
5. $2n - 5 = 9 - 4n$.
6. $V = \frac{1}{3}\pi r^2 h$.
7. $A \geq 21$.
8. $66 < h < 80$.
9. $A = \frac{1}{2}h(b_1 + b_2)$.
10. $A = p + prt$.
11. $2x + 3y = xy$.
12. $\frac{x}{y} \leq 12$.
13. $4(2000)(16) + 5(16) + 3 = 128{,}083$.
14. a. $543 \times 100 = 54{,}300$ cm; b. $\frac{543}{1000} = .543$ km.
15. $4x + 7x + 9x = 20x = 160$; $x = 8$; so the pieces in inches are $4(8) = 32$, $7(8) = 56$, and $9(8) = 72$.
16. $\frac{3(40) + 6(60)}{9} = \frac{480}{9} = \frac{160}{3} = 53\frac{1}{3}$ mph.
17. $A = 2{,}000 + 2{,}000(.08)(10) = \$3{,}600$.
18. $x + (x + 10) = 84$; $x = 37$ and $x + 10 = 47$.
19. $6 + 8x \leq 100$; solving we get $x \leq 11\frac{3}{4}$. The answer is 11 because you can't buy three-fourths of an item.
20. $90 \leq \frac{88 + 95 + 99 + x}{4} \leq 95$; so $360 \leq 282 + x \leq 380$; and $78 \leq x \leq 98$.
21. $\frac{1}{5} + \frac{1}{3} = \frac{3}{15} + \frac{5}{15} = \frac{8}{15}$. A whole job is 1; $1 - \frac{8}{15} = \frac{15}{15} - \frac{8}{15} = \frac{7}{15}$.
22. $\frac{1}{6} + \frac{1}{3} + \frac{1}{2} = \frac{1}{6} + \frac{2}{6} + \frac{3}{6} = \frac{6}{6} = 1$; there is nothing left to do!

Answer to "Bob Asks": Bookkeeper

CHAPTER 8: *Two or More Unknowns*

"Can you name two English words with UU in them?"

SOLVING SIMULTANEOUS EQUATIONS

There are three ways to solve two equations in two unknowns that you may have seen before and two that you probably haven't. There is solving by graphing. We will do that method in the next chapter. The second is by substitution, which we will go over perhaps a little more than you've previously seen. The third is the elimination method, sometimes called the adding and subtracting method. The two that you probably have not seen involve matrices and determinants. The one by matrices, fortunately, you do not need. The second, by determinants, is related to matrices and a great way to solve two equations in two unknowns. We shall see this in a later chapter.

Let's get started.

Substitution

In substitution, we find an unknown with a coefficient of 1, solve for that variable, and substitute it in the other equation.

Example 1: Solve for x and y:

$$3x + 4y = 4 \qquad (1)$$

$$x - 5y = 14 \qquad (2)$$

Solution: In equation (2) $x = 5y + 14$. Substituting this into equation (1), we get $3(5y + 14) + 4y = 4$. Solving, we get $y = -2$; then $x = 5y + 14 = 5(-2) + 14 = 4$. The answer is $x = 4$ and $y = -2$.

We can check by substituting these values into the original equations to see that they are solutions to both equations.

Note *If the coefficient of none of the terms is 1, the elimination method is better.*

Elimination

There are several ways to eliminate one of the variables, as seen in the following examples. Once we have eliminated one of the variables, we can solve for the other variable by using substitution.

Example 2: $2x + 3y = 12$

$5x - 3y = 9$

Solution: If we add the equations, term by term, we can eliminate the y term. We get $7x = 21$, so $x = 3$; substituting $x = 3$ into either equation, we get $y = 2$. So the answer is $x = 3, y = 2$.

If addition doesn't work, try subtraction.

Example 3: $5x + 4y = 14$

$5x - 2y = 8$

Solution: By subtracting, we get $6y = 6$, so $y = 1$; then by substituting, we get $x = 2$.

If adding or subtracting doesn't work, we must find two numbers that, when we multiply the first equation by one of them and the second equation by the other, and then add (or subtract) the resulting equations, one letter is eliminated.

Example 4: $5x + 3y = 11$ (1)

$4x - 2y = 22$ (2)

Solution: To eliminate x, multiply equation (1) by 4 and equation (2) by -5; then add.

$$4(5x + 3y) = 4(11) \quad \text{or} \quad 20x + 12y = 44$$

$$-5(4x - 2y) = -5(22) \quad \text{or} \quad -20x + 10y = -110$$

Adding, we get $22y = -66$; so $y = -3$. We could substitute now, but we could also eliminate y by multiplying the original equation (1) by 2 and equation (2) by 3. Let's do that.

$$2(5x + 3y) = 2(11) \quad \text{or} \quad 10x + 6y = 22$$

$$3(4x - 2y) = 3(22) \quad \text{or} \quad 12x - 6y = 66$$

Adding, we get $22x = 88$, so $x = 4$. The answer is $x = 4, y = -3$.

SOLVING TWO NONLINEAR EQUATIONS IN TWO UNKNOWNS

Sometimes we need to solve two equations in two unknowns when one or both equations are nonlinear (not straight lines).

The nonlinear equations are parabolas, which we will deal with later; circles, which occur throughout this book in different ways; and ellipses and hyperbolas, which are covered later in the book. Collectively, these curves are called **conic sections** because they are curves formed by a plane cutting a cone. If we have one line and one conic section, there are two or fewer (and possibly no) solutions. If we have two conic sections, there are four or fewer solutions.

Example 5: Solve for all points (x, y) in common:

$$y = x^2$$

$$x + y = 12$$

Solution: We shall solve these equations by substitution. We will substitute $y = x^2$ into the second equation.

$x + x^2 = 12$, or $x^2 + x - 12 = (x + 4)(x - 3) = 0$; so $x = -4$ or $x = 3$.

If $x = 3$, then $y = x^2 = 9$, or $3 + y = 12$, and $y = 9$ again; so we have the point $(3, 9)$.

If $x = -4$, $y = 16$; we have the point $(-4, 16)$.

Example 6: Solve for all points (x, y) in common:

$5x^2 + 7y^2 = 29$ (ellipse)

$4x^2 - 5y^2 = 2$ (hyperbola)

Solution: We will use the elimination method here by multiplying the top equation by 4 and the second by -5 and adding:

$$20x^2 + 28y^2 = 116$$

$$-20x^2 + 25y^2 = -10$$

$$\text{so } 53y^2 = 106$$

Then $y^2 = 2$; and $y = \pm\sqrt{2}$. We now substitute either of those values into one of the original equations. Let's take the first one.

$$5x^2 + 7(2) = 29;\ 5x^2 = 15;\ x^2 = 3;\ x = \pm\sqrt{3}.$$

So for each of the y values, there are two answers; this means there are four points in all: $(+\sqrt{3}, +\sqrt{2})$, $(+\sqrt{3}, -\sqrt{2})$, $(-\sqrt{3}, +\sqrt{2})$, $(-\sqrt{3}, -\sqrt{2})$.

Example 7: Solve the pairs of equations:

Problem	Solution
a. $x - y = 20$ $x + y = 10$	$x = 15, y = -5$ $(15, -5)$
b. $5x - y = 23$ $x - y = 7$	$(4, -3)$
c. $4x + 3y = 24$ $2x - 5y = -14$	$(3, 4)$
d. $-3x + 5y = 4$ $-2x + y = 5$	$(-3, -1)$
e. $y = x^2 + 1$ $x + y = 3$	$(1, 2), (-2, 5)$
f. $x^2 + y^2 = 25$ $5x^2 - 3y^2 = 53$	$(4, 3), (4, -3), (-4, 3), (-4, -3)$

Writing Expressions, Equations, and Inequalities in Two or More Unknowns

Sometimes the ACCUPLACER asks you to write expressions or solve equations in two or more unknowns. Before we solve, let's write some of these.

Example 8: Write the following expressions, inequalities, or simultaneous equations:

a. Kira bought *c* canvasses at \$20 each and *t* tubes of oils at \$5 each. Her total cost?

b. Evan bought *c* calculators at \$100 each and special pens *p* at \$12 each. The cost was at least \$500.

c. Sean bought 2 pianos at *p* dollars and 20 pieces of sheet music at *r* dollars each. His cost?

d. Sarah bought *s* soccer balls at \$20 each, *p* pairs of cleats at \$50 each, and *x* soccer shirts at \$25 each. Her total cost?

e. Sheryl, Eric, Glenn, and Wanda pooled their resources and found they had o assets at h dollars each, m assets and y dollars each, w assets at o dollars each, and r debts at d dollars each. Write an expression for their total worth.

f. Marlene found that if she bought 5 hamburgers at x dollars each and 4 hot dogs at y dollars each, she would pay \$41, while 3 hamburgers and 7 hotdogs at the same price would cost \$43. Write the equations to describe this.

Solutions:

a. $20c + 5t$

b. $100c + 12p \geq 500$

c. $2p + 20r$

d. $20s + 50p + 25x$

e. $oh + my + wo - rd$

f. $5x + 4y = 41$
$3x + 7y = 43$

Let's do some multiple-choice exercises.

Exercise 1: If $x = y + 3$ and $y = z + 7$, x (in terms of z) =

A. $z - 10$
B. $z - 4$
C. z
D. $z + 4$
E. $z + 10$

Exercise 2: Two apples and 3 pears cost 65 cents, and 5 apples and 4 pears cost \$1.10. Find the cost, in cents, of one pear:

A. 10
B. 15
C. 20
D. 25
E. 30

Exercise 3: As in Exercise 2, 2 apples and 3 pears cost 65 cents, and 5 apples and 4 pears cost \$1.10. Find the cost, in cents, of one pear and one apple together:

A. 10
B. 15
C. 20
D. 25
E. 30

Exercise 4: Find $x + y$:

$$7x + 4y = 27$$
$$x - 2y = -3$$

A. 1
B. 3
C. 5
D. 7
E. 9

Exercise 5: For lunch, Ed buys 3 hamburgers and 1 soda for \$12.50, and Mei buys 1 hamburger and 1 soda for \$5.60. How much does Ed pay for his hamburgers?

A. \$2.15
B. \$3.45
C. \$6.90
D. \$10.35
E. \$18.10

Exercise 6: The product of 4 and the sum of x and y is at least as large as the quotient of a divided by b. This can be written as

A. $4x + y - \frac{a}{b} \geq 0$
B. $x + 4y - \frac{a}{b} > 0$
C. $4(x + y) - \frac{a}{b} \geq 0$
D. $\frac{a}{b} - 4x + 4y < 0$
E. $4(x + y) + \frac{a}{b} \geq 0$

Let's look at the answers.

Answer 1: **E:** $x = y + 3 = (z + 7) + 3 = z + 10$.

Answer 2: **B:**

$2a + 3p = 65$

$5a + 4p = 110$

In solving for p, eliminate a by multiplying the top equation by 5 and the bottom by -2.

$5(2a + 3p) = 5(65)$ or $10a + 15p = 325$

$-2(5a + 4p) = -2(110)$ or $-10a - 8p = -220$

Adding, we get $7p = 105$; $p = 15$.

Answer 3: **D:** Much more often, we get a problem like this. The equations are the same as in Exercise 2, but rather than asking for the cost of one apple or the cost of one pear, this exercise asks for the cost of one apple plus one pear. The trick is simply to add the original equations. We then get $7a + 7p = 175$. Dividing both sides by 7, we get $a + p = 25$.

Answer 4: **C:** Less frequently, when adding doesn't work, try subtracting. If we subtract, the difference becomes $6x + 6y = 30$. So $x + y = 5$.

Answer 5: **D:** The equations are

$3h + s = 12.50$

$h + s = 5.60$

Subtracting, we get $2h = 6.90$, so $h = 3.45$. Ed's three hamburgers cost \$10.35.

Answer 6: **C:** $4(x + y) \geq \frac{a}{b}$. After rearranging, only answer choice C is correct.

Chapter 8 Test

For questions 1–14, solve for x and y:

1. $x - y = 12$

   $x + y = 2$

2. $4x - 3y = 25$

   $x - 3y = 10$

3. $2x - 5y = -6$

   $5x - 7y = -15$

4. $4x + 6y = 3$

   $8x - 9y = -1$

5. $\frac{x}{4} - \frac{y}{3} = 2$

   $\frac{x}{6} - \frac{5y}{3} = -3$

6. $y = 3x + 2$

   $4x + 3y = 32$

7. $ax + by = f$

   $cx + dy = g$

8. $y = x^2 - 4$

   $y = 4x + 1$

9. $y = x^2 + 1$

   $x^2 + y^2 = 1$

10. $2x^2 + 3y^2 = 12$

    $5x^2 - 4y^2 = 7$

11. $x = y^2 + 2y + 3$

    $x = 3y^2 + 8y + 7$

12. $xy = 1$

    $x^2 + y^2 = 2$

13. $4x - 6y = 12$

    $6x - 9y = 18$

14. $2x - 7y = 3$

    $-6x + 21y = 5$

15. Sue bought x dresses at \$70 and y shoes at \$50 a pair. She has at most \$1,000 to spend. Write this as an inequality.

16. The amount of x \$10 bills, y \$50 bills, and z \$100 bills totals more than \$10,000. Write the inequality.

17. The cost of 4 hamburgers and 6 hot dogs is \$29. The cost of 7 hamburgers and 3 hot dogs is \$32. Find the cost of one hamburger and one hot dog.

18. I have 30 dimes and nickels that total \$2.40. How many of each coin do I have?

19. Four times of sum of x and y minus three times the difference of x and y is 19. The sum of the numbers is 7. Find the numbers.

20. At a modern brand new major league ballpark, believe it or not, the cost of 12 deluxe box seats and 5 "ordinary" box seats cost \$11,450 (for one game!!!!), and the cost of 7 deluxe box seats and 3 "ordinary" box seats cost \$6700. Find the cost of one deluxe and one ordinary box seat.

21. Solve for x and y: $2x - 5y = 10$

    $4x - 10y = 30$

22. Solve for x and y: $3x + 2y = 7$

    $9x + 6y = 21$

Answers to Chapter 8 Test

1. Add them; $2x = 14$; $x = 7$; then substitute: $(7, -5)$.

2. Subtract; $3x = 15$; $x = 5$; substitute $\left(5, -\frac{5}{3}\right)$.

3. Multiply the top equation by 5 and the bottom equation by -2; $-11y = 0$; $y = 0$; $(-3, 0)$.

4. Multiply the top equation by -2 and add; $-21y = -7$; $y = \frac{1}{3}$; $\left(\frac{1}{4}, \frac{1}{3}\right)$.

5. First, multiply the top equation by 12 and the bottom equation by 6 to get rid of the fractions; we then have $3x - 4y = 24$ and $x - 10y = -18$. Multiply the second equation by -3 and add; $26y = 78$; $y = 3$; $(12, 3)$.

6. Use substitution and add; $4x + 3(3x + 2) = 32$; $13x + 6 = 32$; $x = 2$; $(2, 8)$.

7. For y, multiply the top equation by c and the bottom by $-a$; $y = \frac{ag - cf}{ad - bc}$; for x, multiply the top equation by d and the bottom by $-b$ and add; $x = \frac{df - bg}{ad - bc}$.

8. $x^2 - 4 = 4x + 1$; $x^2 - 4x - 5 = (x - 5)(x + 1) = 0$; $(5, 21)$ and $(-1, -3)$.

9. This is a little twist; let $x^2 = y - 1$; so $y - 1 + y^2 = 1$, or $y^2 + y - 2 = (y + 2)(y - 1) = 0$; so $y = -2$ or $y = 1$; if $y = 1$; then $x = 0$; if $y = -2$; $-2 = x^2 + 1$; $x^2 = -3$; so there is no real value for x; this problem has only one solution, $(0, 1)$.

10. Multiply the top equation by 4 and the bottom equation by 3; $23x^2 = 69$; $x^2 = 3$; $x = \pm\sqrt{3}$; substituting each value, we get $\pm\sqrt{2}$; for each one; the four answers are $(\sqrt{3}, \sqrt{2})$, $(\sqrt{3}, -\sqrt{2})$, $(-\sqrt{3}, \sqrt{2})$, $(-\sqrt{3}, -\sqrt{2})$.

11. $3y^2 + 8y + 7 = y^2 + 2y + 3$; $2y^2 + 6y + 4 = 2(y + 1)(y + 2) = 0$; $y = -2$ or -1; $(2, -1)$, $(3, -2)$.

12. $y = \frac{1}{x}$; $x^2 + \left(\frac{1}{x}\right)^2 = 2$; multiply by x^2 and rearrange highest exponent to lowest. You get

$x^4 - 2x^2 + 1 = (x^2 - 1)(x^2 - 1) = (x + 1)(x - 1)(x + 1)(x - 1) = 0$; $x = 1$ or -1; always substitute in the easier equation, and you will not get extra (extraneous) nonanswers; $(1, 1), (-1, -1)$.

13. Multiply the top equation by 3 and the bottom equation by -2; after you add, you will get $0 = 0$, which means there isn't just one answer. Both equations are the same, so any answer to one is an answer to the other. You may want to look at this one again after the next chapter (you can see the answer by graphing carefully).
14. If you multiply the top equation by 3 and then add the equations, you get $0 = 14$, which is never true. This means the two equations have no common points. If you graph the equations, the lines are parallel.
15. $70x + 50y \leq \$1{,}000$.
16. $10x + 50y + 100z > 10{,}000$.
17. $4H + 6D = 29$ and $7H + 3D$ (this problem has depth) $= 32$; multiply the bottom equation by -2 and add them; $-10H = -35$; $H = \$3.50$; $D = \$2.50$. So the cost is \$6.00.
18. $N + D = 30$; $5N + 10D = 240$; $5(30 - D) + 10D = 240$; $150 - 5D + 10D = 240$; $5D = 90$; $D = 18$; $N = 12$. Check? $18(10) + 12(5) = 240$ pennies.
19. $4(x + y) - 3(x - y) = x + 7y = 19$ and $x + y = 7$; so $7 - y + 7y = 19$; $6y = 12$; $y = 2$; $x = 5$
20. Let D = deluxe and O = ordinary (OK, you shouldn't use O, but I will); $12D + 5O = 11{,}450$, and $7D + 3O = 6700$; multiply the first equation by 3 and the second by -5; $36D + 15O = 34{,}350$, and $-35D - 15O = -33{,}500$; $D = \$850.00$! and $O = \$250$! Take me out to the ball game! After this question, I'm ready for anything!
21. There are no values for x and y, no solution. Multiplying the top equation by -2 and adding, you would get $0 = 20$ which, of course, is not true. If you graphed these lines, you would see they are parallel.
22. An infinite number of points, the whole line, is the answer. If you multiplied the top equation by -3 and added, you would get $0 = 0$, which is always true. In graphing linear equations in two unknowns on the x-y plane, there are all three possibilities: one point (the two lines intersect), no points(the lines are parallel), and an infinite number of points (the line itself—both equations—are the same line).

Let's take a break from algebra, and take a look at points, lines, and shapes.

Answer to "Bob Asks": Vacuum and continuum

CHAPTER 9: *Points, Lines, and More*

"A clock rings the number of hours each hour and it rings once on the half-hour. At six o'clock the bell rings six times; at 6:30 once, and at seven o'clock the bell rings seven times. John comes home to hear the bell ring once. A half-hour later it rings once, a half-hour later it rings once again, and a half-hour later, it rings once again. If the clock is not broken, when did John come home?"

This topic used to be part of a course called analytic geometry (algebraic geometry). We'll start at the beginning.

POINTS IN THE PLANE

We start with a **plane**—a two-dimensional space, like a piece of paper. On this plane, we draw two perpendicular lines, or **axes**. The *x*-axis is horizontal; the *y*-axis is vertical. Positive *x* is to the right; negative *x* is to the left. Positive *y* is up; negative *y* is down. Points in the plane are indicated by **ordered pairs** (*x*, *y*). The *x* number, called the **first coordinate** or **abscissa**, is always given first; the *y* number, called the **second coordinate** or **ordinate**, is always given second. Here are some points on the plane.

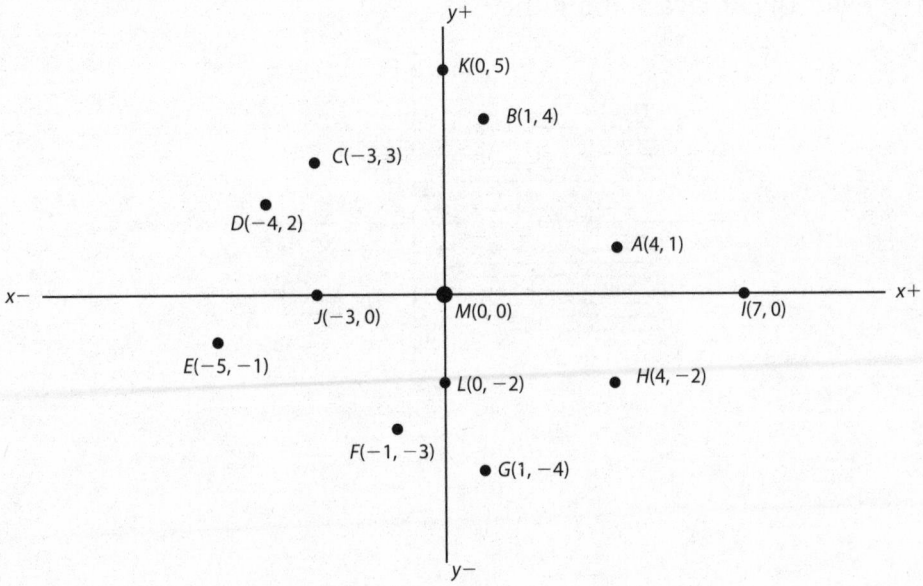

Note the following:

- For any point on the x-axis, the y-coordinate always is 0.
- For any point on the y-axis, the x-coordinate is 0.
- The point where the two axes meet, (0, 0), is called the **origin**.
- The axes divide the plane into four **quadrants**, usually written with roman numerals, starting in the upper right quadrant and going counterclockwise.
  - In quadrant I, $x > 0$ and $y > 0$.
  - In quadrant II, $x < 0$ and $y > 0$.
  - In quadrant III, $x < 0$ and $y < 0$.
  - In quadrant IV, $x > 0$ and $y < 0$.

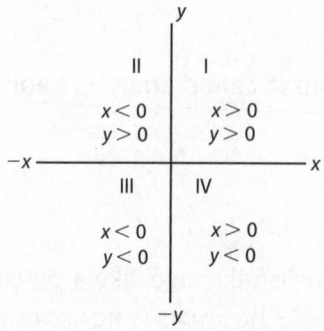

In the following figure, we have drawn the line $y = x$. For every point on this line, the first coordinate has the same value as the second coordinate, or $y = x$. If we shade the area above this line, $y > x$ in the shaded portion. Similarly, $x > y$ in the unshaded portion. Sometimes, questions on the ACCUPLACER ask about this.

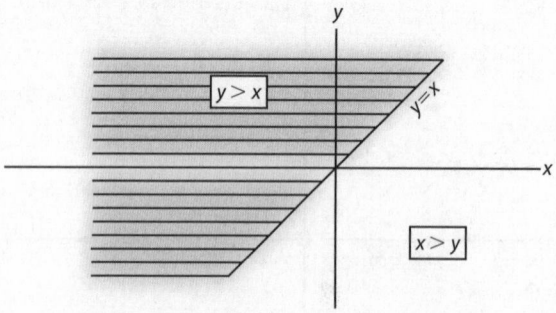

The following figure shows symmetry of the point (a, b) about the x-axis, y-axis, and the origin. Suppose (a, b) is in quadrant I. Then $(-a, b)$ would be in quadrant II, $(-a, -b)$ would be in quadrant III, and $(a, -b)$ would be in quadrant IV, as pictured.

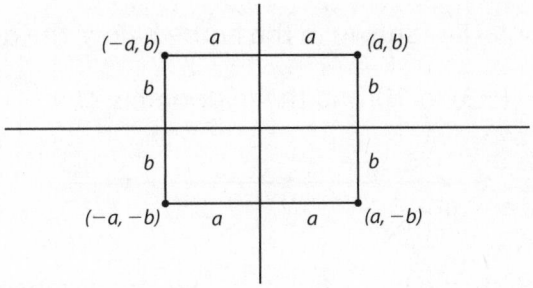

LINES

Distance and Midpoint

The formulas for distance and midpoint look a little complicated, but they are fairly easy to use. It just takes practice.

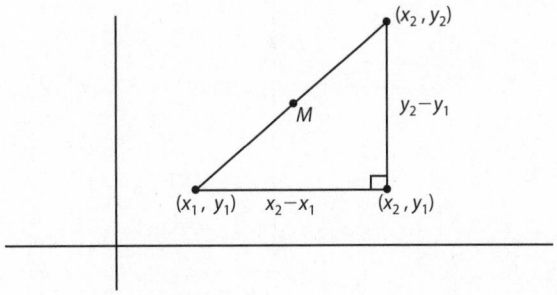

To find the **distance** between two points (x_1, y_1) and (x_2, y_2) on a plane, we must use the **distance formula**:

$$d = \sqrt{(x_2 - x_1)^2 + (y_2 - y_1)^2}$$

The distance formula is just the Pythagorean Theorem (discussed in the next chapter).

Distances are always positive. You may be six feet tall, but you cannot be minus six feet tall.

The **midpoint** of a line between two points (x_1, y_1) and (x_2, y_2) on a plane is given by the coordinates

$$M = \left(\frac{x_1 + x_2}{2}, \frac{y_1 + y_2}{2}\right)$$

In one dimension, if the line is horizontal, these formulas simplify to $d = |x_2 - x_1|$ and $M = \frac{x_1 + x_2}{2}$.

Similarly, if the line is vertical, these formulas simplify to $d = |y_2 - y_1|$ and $M = \frac{y_1 + y_2}{2}$.

For example, for the horizontal line shown in the figure below, the distance between the points is $d = |x_2 - x_1| = |7 - (-3)| = 10$, and the midpoint is $M = \frac{x_1 + x_2}{2} = \frac{-3 + 7}{2} = 2$.

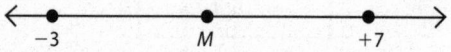

Similarly, for the vertical line shown in the figure below, the distance between the points is $d = |y_2 - y_1| = |-4 - (-8)| = 4$, and the midpoint is $M = \frac{y_1 + y_2}{2} = \frac{(-8) + (-4)}{2} = -6$.

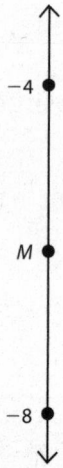

Slope

The **slope** of a line tells by how much the line is "tilted" compared to the x-axis. The formula for the slope of a line is

$$m = \frac{\text{rise}}{\text{run}} = \frac{\text{change in } y}{\text{change in } x} = \frac{y_2 - y_1}{x_2 - x_1},$$

where (x_1, y_1) and (x_2, y_2) are any two points on the line.

Note the following facts about the slope of a line, as shown in the figure below:

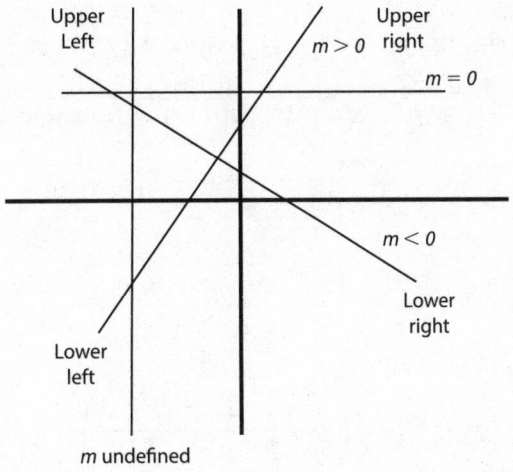

- The slope is positive if the line goes from the lower left to the upper right.
- The slope is negative if it goes from the upper left to the lower right.
- Horizontal lines have zero slope.
- Vertical lines have no slope or undefined slope or "infinite" slope.

Example 1: Find the distance, slope, and midpoint for the line segment joining these points:

a. (2, 3) and (6, 8)

b. (4, −3) and (−2, 0)

c. (7, 3) and (4, 3)

d. (2, 1) and (2, 5)

Solutions:

a. We let $(x_1, y_1) = (2, 3)$ and $(x_2, y_2) = (6, 8)$, although the other way around is also okay.

Then

$$\text{Distance} = d = \sqrt{(x_2 - x_1)^2 + (y_2 - y_1)^2} = \sqrt{(6-2)^2 + (8-3)^2} = \sqrt{41}$$

$$\text{Slope} = m = \frac{y_2 - y_1}{x_2 - x_1} = \frac{8-3}{6-2} = \frac{5}{4}$$

$$\text{Midpoint} = M = \left(\frac{x_1 + x_2}{2}, \frac{y_1 + y_2}{2}\right) = \left(\frac{2+6}{2}, \frac{3+8}{2}\right) = (4, 5.5)$$

Notice that the slope is positive; the line segment goes from the lower left to the upper right.

b. We let $(x_1, y_1) = (4, -3) =$ and $(x_2, y_2) = (-2, 0)$.

$$\text{Distance} = d = \sqrt{(-2-4)^2 + (0-(-3))^2} = \sqrt{45}$$

$$= \sqrt{3 \times 3 \times 5} = 3\sqrt{5}$$

$$\text{Slope} = m = \frac{0-(-3)}{-2-4} = -\frac{1}{2}$$

$$\text{Midpoint} = M = \left(\frac{4+(-2)}{2}, \frac{-3+0}{2}\right) = (1, -1.5)$$

Notice that the slope is negative; the line segment goes from the upper left to the lower right.

c. We let $(x_1, y_1) = (7, 3) =$ and $(x_2, y_2) = (4, 3)$.

It is a one-dimensional horizontal distance (the y values are the same), so $d = |4 - 7| = 3$

$$\text{Slope} = m = \frac{3-3}{4-7} = \frac{0}{-3} = 0$$

$$\text{Midpoint} = M = \left(\frac{7+4}{2}, \frac{3+3}{2}\right) = (5.5, 3)$$

Notice that the horizontal line segment has slope $m = 0$.

d. We let $(x_1, y_1) = (2, 1) =$ and $(x_2, y_2) = (2, 5)$.

Again, this is a one-dimensional distance (but vertical since the x values are the same), so $d = |5 - 1| = 4$

$$\text{Slope} = m = \frac{5-1}{2-2} = \frac{4}{0}, \text{ undefined (cannot divide by 0)}$$

$$\text{Midpoint} = M = \left(\frac{2+2}{2}, \frac{5+1}{2}\right) = (2, 3)$$

Notice that the slope of the vertical line segment is undefined.

Now let's do some multiple-choice exercises.

Exercise 1: The coordinates of P are (j, k). If $s < k < j < r$, which of the points shown in the figure could have the coordinates (r, s)?

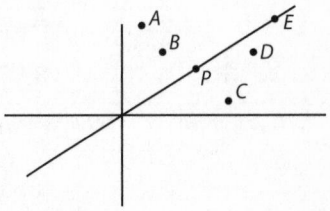

A. A

B. B

C. C

D. D

E. E

Use the figure below for Exercises 2 and 3.

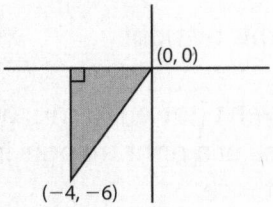

Exercise 2: Which of the following points is inside the triangle?

A. $(-3, 6)$

B. $(-5, -5)$

C. $(-2, -5)$

D. $(-3, -4)$

E. $(-1, -3)$

Exercise 3: The area of the triangle is

A. 6

B. 12

C. 18

D. 24

E. 48

Exercise 4: M is the midpoint of the horizontal line segment AB. If the coordinates of A are $(m, -n)$, then the coordinates of B are

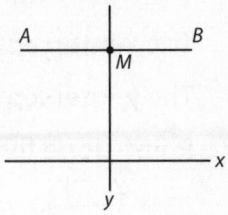

A. (m, n)

B. $(-m, n)$

C. $(-m, -n)$

D. (n, m)

E. $(-n, -m)$

Exercise 5: In the given figure, AB is parallel to ($\parallel$) the x-axis and $PQ = AB$. The coordinates of point A are

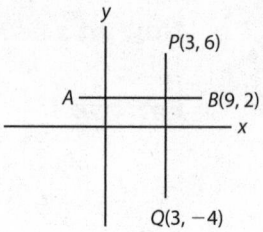

A. $(-1, 2)$

B. $(1, 2)$

C. $(9, -8)$

D. $(9, 0)$

E. $(-9, 2)$

Let's look at the answers.

Answer 1: **C:** For points C, D, and E, the x value is bigger than the x value of P; of these, only point C has a y value less than the y value of P.

Answer 2: **D:** You can do this by sight.

Answer 3: **B:** We really haven't gotten to this, but I asked it because we have the picture. The area of a right triangle is half the horizontal distance times the vertical distance.

$$A = \frac{1}{2}bh = \frac{1}{2} \times 4 \times 6 = 12$$

Answer 4: **C:** Slightly tricky. Point B has the same y value as A, but its x value is the negative of the x value for A. Note that the actual values for m and n must be negative numbers in this figure.

Answer 5: **A:** The length of $PQ = 10$. For the length of AB to be 10, A must be $(-1, 2)$ since $|9 - (-1)| = 10$.

Standard Equation of a Line

Let's go over the facts we need.

- **Standard form** of the line: $Ax + By = C$; A, B both $\neq 0$.
- The ***x*-intercept**, the point at which the line hits the x-axis, occurs when $y = 0$.
- The ***y*-intercept**, the point at which the line hits the y-axis, occurs when $x = 0$.
- **Point-slope form** of a line: Given slope m and point (x_1, y_1), the point-slope form of a line is $m = \dfrac{y - y_1}{x - x_1}$.

- **Slope-intercept form** of a line: $y = mx + b$, where m is the slope and $(0, b)$ is the y-intercept.
- Lines of the form:

  $y =$ constant, such as $y = 2$, are lines parallel to the x-axis; the equation of the x-axis is $y = 0$.

  $x =$ constant, such as $x = -3$, are lines parallel to the y-axis; the equation of the y-axis is $x = 0$.

  $y = mx$ are lines that pass through the origin.

Example 2: For $Ax + By = C$, find the x and y intercepts.

Solution: The y-intercept means $x = 0$; so $y = \frac{C}{B}$, and the y-intercept is $\left(0, \frac{C}{B}\right)$.

The x-intercept means $y = 0$; so $x = \frac{C}{A}$, and the x-intercept is $\left(\frac{C}{A}, 0\right)$.

Example 3: For $3x - 4y = 7$, find the x and y intercepts.

Solution: For the y-intercept, $x = 0$; so $y = \frac{7}{-4}$, and the y-intercept is $\left(0, -\frac{7}{4}\right)$.

For the x-intercept, $y = 0$; so $x = \frac{7}{3}$, and the x-intercept is $\left(\frac{7}{3}, 0\right)$.

Example 4: Given a line with $m = \frac{3}{2}$ and containing point $(5, -7)$, write the equation of the line in standard form.

Solution: $m = \frac{y - y_1}{x - x_1}$, so $\frac{3}{2} = \frac{y - (-7)}{x - 5}$. Cross-multiplying, we get $3(x - 5) = 2(y + 7)$, or $3x - 2y = 29$.

Example 5: Given points $(3, 6)$ and $(7, 11)$ on a line, write the equation of the line in slope-intercept form.

Solution: $y = mx + b$. $m = \frac{11 - 6}{7 - 3} = \frac{5}{4}$, and we will use point $(3, 6)$, so $x = 3$ and $y = 6$. Therefore, $6 = \frac{5}{4}(3) + b$, and $b = \frac{9}{4}$. So the line is $y = \frac{5}{4}x + \frac{9}{4}$.

Example 6: Sketch lines $x = -3$, $y = 8$, and $y = \frac{2}{3}x$.

Solution:

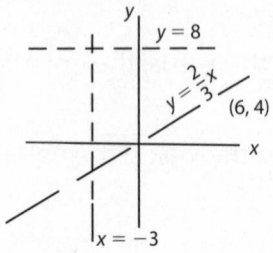

Q **Let's do a couple more multiple-choice exercises.**

Exercise 6: A line with the same slope as the line $y = \frac{2}{3}x - 5$ is

A. $2x = 18 - 3y$

B. $2x + 3y = 6$

C. $2x - 3y = 6$

D. $-2x - 3y = 14$

E. $2y = 12 - 3x$

Exercise 7: Find the area of the triangle formed with the positive x-axis, positive y-axis, and the line through the point (3, 4) with slope -2. The area is

A. 5

B. 15

C. 25

D. 50

E. 10

A **Let's look at the answers.**

Answer 6: **C:** We have to solve for y in each case, but we are interested in only the coefficient of x. The only answer choice that works is *C*.

Answer 7: **C:** You must visualize the figure.

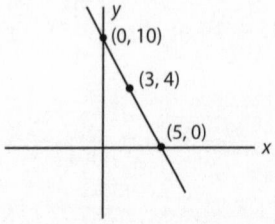

Using the point-slope form, the equation of the line is $-2 = \frac{y-4}{x-3}$.
If we let $x = 0$, the y-intercept is 10. If we let $y = 0$, the x-intercept is 5.
The area of the triangle is one-half the x distance times the y distance.
Area $= \frac{1}{2}ab = \frac{1}{2} \times 5 \times 10 = 25$.

Parallel and Perpendicular Lines

We say two lines are **parallel** if they have the **same slope**.

Note *There is a special case that two lines are parallel also if they are parallel to the y-axis. We need this case since lines parallel to the y-axis are said to have infinite slope. Since infinity is not a number, we can't say infinity equals infinity. In fact, as a teaser for those who care, there is more than one infinity. In fact, there are an infinite number of infinities. Let's get on with it!*

We say two lines are **perpendicular** if their slopes are **negative reciprocals** of each other. Again, there is a special case for this. Two lines are also perpendicular if one is parallel to the x-axis (slope 0) and one is parallel to the y-axis (infinite slope).

Example 7: Given line L: $7x + 8y = 9$.

a. Find the equation in standard form of a line parallel to L through the point (2, 3).

b. Find the equation in standard form of a line perpendicular to L through the point (5, −6).

Solution: In each case we must find the slope. Solving for y, we get $y = -\frac{7}{8}x + \frac{9}{8}$.

a. Since the slope is the coefficient of x, $-\frac{7}{8}$, the parallel line will have the same slope, $-\frac{7}{8}$, and the point (2, 3). We can use the point-slope form: $m = \frac{y - y_1}{x - x_1}$, or $\frac{-7}{8} = \frac{y-3}{x-2}$. Cross-multiplying, we get $-7(x - 2) = 8(y - 3)$; $-7x + 14 = 8y - 24$; rearranging we get $7x + 8y = 38$.

b. The perpendicular line will have a slope that is the negative reciprocal of $\frac{-7}{8}$, or $+\frac{8}{7}$. The point-slope form is $m = \frac{y - y_1}{x - x_1}$, or $\frac{8}{7} = \frac{y-(-6)}{x-5}$. Then $8(x - 5) = 7(y + 6)$, or $8x - 7y = 82$.

Solving Two Equations in Two Unknowns by Graphing

It may be necessary to solve two equations in two unknowns by graphing. What you must do is graph each line. The point where they meet is the solution to the problem.

Example 8: Solve by graphing: $2x + y = 8$
$2x + 3y = 12$

Solution: We graph by using intercepts.

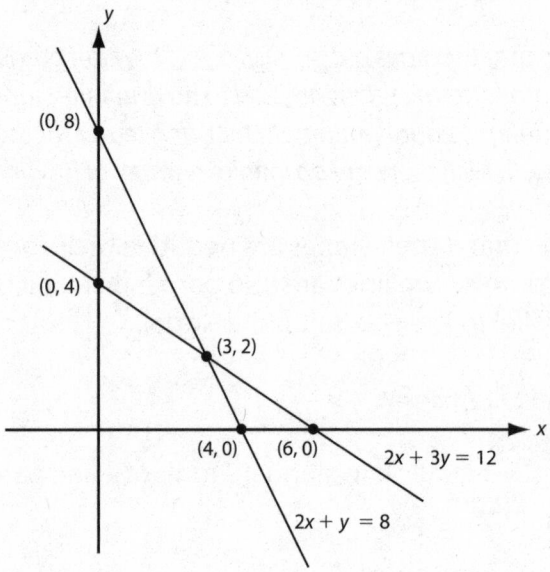

$2x + y = 8$: If $x = 0$, then $y = 8$, so the y-intercept is (0, 8); if $y = 0$, then $x = 4$, so the x-intercept is (4, 0). Graph the line through these two points.

$2x + 3y = 12$: If $x = 0$, then $y = 4$, and the y-intercept is (0, 4); if $y = 0$, then $x = 6$, and the x-intercept is (6, 0). Graph the line through these two points. The lines meet at the point $P(3, 2)$. You can check $x = 3$ and $y = 2$ in both equations. On the ACCUPLACER, this method will rarely be used since drawing pictures would lead to inaccuracies and take more time.

EQUATION OF A CIRCLE

Here we do an algebraic version of the circle. We will do more, like in your geometry class, in Chapter 12.

A circle is the set of all points (x, y) at a given distance r (called the **radius**) from a given point (h, k) called the center.

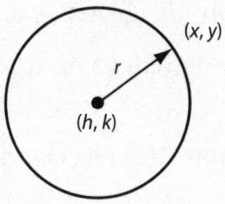

Finding the radius means using the distance formula, so we get the following:

$$r = \sqrt{(x - h)^2 + (y - k)^2}$$

Note *(h, k) is the center, a fixed point.*

Note *(x, y) is any point on the circumference.*

No one uses this form of the circle. Instead, we square both sides. So the equation of the circle is $(x - h)^2 + (y - k)^2 = r^2$.

Example 9: What is the center and radius of the circle $(x - 5)^2 + (y + 6)^2 = 11$.

Solution: Using the equation of a circle, we see that $h = 5$, $k = 6$, and $r^2 = 11$. So the center is the point $(5, -6)$ and the radius is $\sqrt{11}$.

Example 10: Write the equation of a circle with radius 7 and center $(-8, 9)$.

Solution: Substituting directly into the equation of a circle, we get $(x + 8)^2 + (y - 9)^2 = 49$.

Note *If you get any questions on these last two topics, especially circles, sketch the picture. You'll be surprised how many times the picture will give you the answer.*

Chapter 9 Test

For questions 1–6, let point A be $(4, -7)$ and point B be $(-4, -11)$.

1. Find the midpoint of AB.
2. Find the slope of AB.
3. Find the length of AB.
4. Find the equation of the line through AB. (Give the answer in standard form.)
5. Find the equation of the line perpendicular to AB through $(-2, 1)$. (Give the answer in point-slope form.)
6. Find the equation of the circle that has AB as its diameter.
7. What is the equation of any line perpendicular to the y-axis?
8. The line $3x - 4y = 5$ goes through which quadrants?
9. Write the equation of the line through $(5, 7)$ parallel to the y-axis.
10. Write the equation of the line with x-intercept 4 and y-intercept 2.
11. Write the equation of a circle with center $(4, -6)$ and radius 3.
12. Find the center and radius of a circle if the equation is $(x - 11)^2 + (y + 4)^2 = 13$.
13. Write the equation of a circle with center at $(-5, -4)$, tangent to (just touching) the y-axis.

For questions 14 and 15, L is $4y - 8x = 27$.

14. Find the equation of a line parallel to L through $(-8, 3)$. (Give the answer in point-slope form.)
15. Find the equation of the line perpendicular to L though $(-5, -7)$. (Give the answer in point-slope form.)

For questions 16–20, let point $M = (a, b)$ and point $N = (5a, 7b)$.

16. Find the midpoint of MN.
17. Find the slope of MN.
18. Find the distance between M and N.
19. Find the equation of the line parallel to MN through the point $(c, -d)$. (Give the answer in point-slope form.)
20. Find the equation of the circle with MN as the diameter.

Answers to Chapter 9 Test

1. $\left(\dfrac{4 + (-4)}{2}, \dfrac{(-7) + (-11)}{2}\right) = (0, -9)$.

2. $m = \frac{y_2 - y_1}{x_2 - x_1} = \frac{-11 - (-7)}{-4 - 4} = \frac{-4}{-8} = \frac{1}{2}$.

3. $\sqrt{(x_2 - x_1)^2 + (y_2 - y_1)^2} = \sqrt{(-4 - 4)^2 + [-11 - (-7)]^2} = \sqrt{(-8)^2 + (-4)^2} = \sqrt{80} = 4\sqrt{5}$.

4. From question 2, the slope is $\frac{1}{2}$. Then $\frac{1}{2} = \frac{y - (-7)}{x - 4}$, or $x - 2y = 18$.

5. For a perpendicular line, the slope is the negative reciprocal, so $-2 = \frac{y - 1}{x - (-2)} = \frac{y - 1}{x + 2}$. On a class test, you must leave it in the form the teacher asks. On the ACCUPLACER, you must match the form of your answer to the test answer choices. Don't do any more work than you have to. It's the way of the mathematician.

6. $r = \frac{1}{2}d = \frac{1}{2}(4\sqrt{5}) = 2\sqrt{5}$. The midpoint is the center (from question 1), $(0, -9)$. The equation of the circle is therefore $x^2 + (y + 9)^2 = 20$.

7. $x =$ any number.

8. I, III, and IV (sketch the picture by using intercepts).

9. $x = 5$.

10. The points are $(4, 0)$ and $(0, 2)$; $m = \frac{2 - 0}{0 - 4} = -\frac{1}{2}$. In slope-intercept form, $y = -\frac{1}{2}x + 2$.

11. $(x - 4)^2 + (y + 6)^2 = 3^2$

12. $(11, -4)$, $r = \sqrt{13}$.

13. $(x + 5)^2 + (y + 4)^2 = (4)^2$.

14. $m = 2$; $2 = \frac{y - 3}{x + 8}$.

15. The slope of L is 2; the slope of a line perpendicular is $-\frac{1}{2}$. The equation is $-\frac{1}{2} = \frac{y + 7}{x + 5}$.

16. $(3a, 4b)$.

17. $\frac{6b}{4a} = \frac{3b}{2a}$.

18. $\sqrt{(5a - a)^2 + (7b - b)^2} = \sqrt{16a^2 + 36b^2} = 2\sqrt{4a^2 + 9b^2}$.

19. From question 17, the slope of MN is $\frac{3b}{2a} = \frac{y + d}{x - c}$. Letters are nice—no arithmetic to do.

20. From question 18, we have $d = 2\sqrt{4a^2 + 9b^2}$, and for the radius, we divide by 2. From question 16, the center is $(3a, 4b)$. So the equation is $(x - 3a)^2 + (y - 4b)^2 = 4a^2 + 9b^2$.

Let's finally get to angles and triangles.

Answer to "Bob Asks": John comes in just in time to hear the last ring of 12:00. This question shows you that you must understand the question before you answer it.

CHAPTER 10: *All About Angles and Triangles*

"Can you name the shortest word in the English language with all five vowels (not necessarily in order)? The word is seven letters long."

Before I wrote this chapter, I formulated in my head how the chapter would go. Too many of the questions on angles had to do with triangles. So I decided to write the chapters together. Let's start with some definitions.

TYPES OF ANGLES

There are several ways to classify angles, such as by angle measure, as shown here:

Acute angle: An angle of less than 90°.

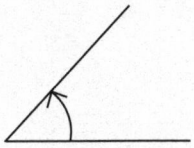

acute

Right angle: A 90° angle. As we will see, some other words that indicate a right angle or angles are perpendicular (⊥), altitude, and height.

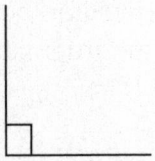

right

Obtuse angle: An angle of more than 90° but less than 180°.

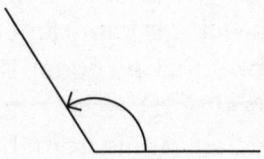

obtuse

Straight angle: An angle of 180°.

Reflex angle: An angle of more than 180° but less than 360°.

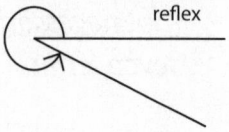

Angles are also named for their relation to other angles, such as:

Supplementary angles: Two angles that total 180°. $\angle 1 + \angle 2 = 180°$.

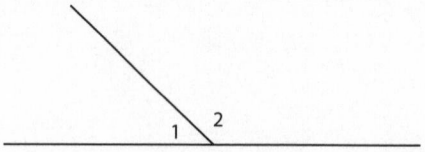

$\angle 1 + \angle 2 = 180°$

Complementary angles: Two angles that total 90°. $\angle 1 + \angle 2 = 90°$.

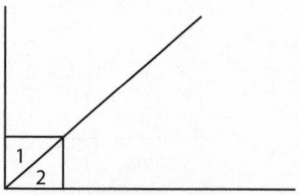

$\angle 1 + \angle 2 = 90°$

You probably learned that angles are congruent and measures of angles are equal. I am using what I learned; it is simpler and makes understanding easier. So "angle 1 equals angle 2" (or $\angle 1 = \angle 2$) means the angles are both congruent and equal in degrees.

ANGLES FORMED BY PARALLEL LINES

Let's look at angles formed when a line crosses two parallel lines. In the next figure, $\ell_1 \| \ell_2$, and t is a **transversal**, a line that cuts two or more lines. It is not important that you know the names of these angles, although many of you will. It is important only to know that angles formed by a line crossing parallel lines that look equal are equal. The angles that are adjacent to these angles add to 180° with them. In this figure, $\angle 1 = \angle 4 = \angle 5 = \angle 8$ and $\angle 2 = \angle 3 = \angle 6 = \angle 7$. Any angle from the first group added to any angle from the second group totals 180°.

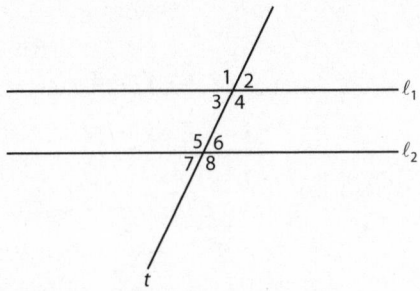

Vertical angles, which are the opposite angles formed when two lines cross, are equal. In the figure below, $\angle 1 = \angle 3$ and $\angle 2 = \angle 4$. Also, $\angle 1 + \angle 2 = \angle 2 + \angle 3 = \angle 3 + \angle 4 = \angle 4 + \angle 1 = 180°$.

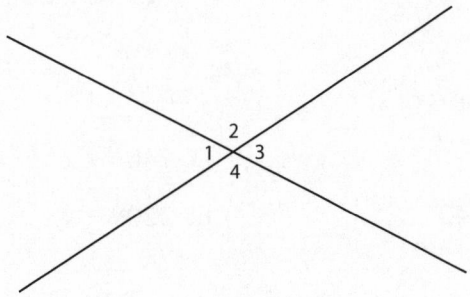

Q **Let's do some multiple-choice exercises.**

Exercise 1: $\angle b =$

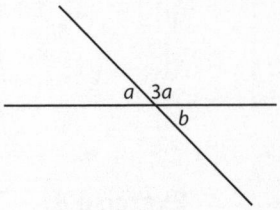

A. 45°

B. 60°

C. 90°

D. 105°

E. 135°

Exercise 2: $\ell_1 \parallel \ell_2$. $m - n =$

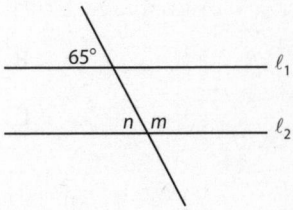

A. 30°

B. 50°

C. 65°

D. 90°

E. 180°

Exercise 3: $y + z =$

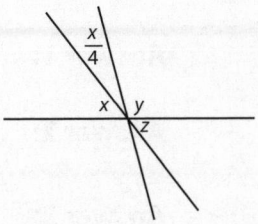

A. $180° - x$

B. $180° - \frac{x}{4}$

C. $45° - \frac{x}{4}$

D. $90° + \frac{5x}{4}$

E. $90° - \frac{5x}{4}$

Exercise 4: $180° - w =$

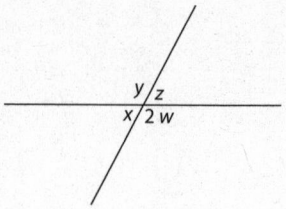

A. $x + w$

B. $x + y$

C. $y + z$

D. $y - z$

E. $z - w$

Exercise 5: $b =$

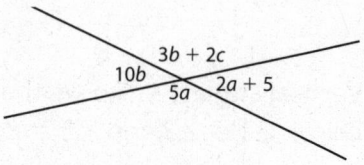

A. 5.5°

B. 7°

C. 10°

D. 12.5°

E. Cannot be determined

Exercise 6: y (in terms of x) =

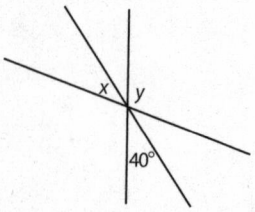

A. x

B. $x + 40°$

C. $140° - x$

D. $140° + x$

E. $320° - x$

Exercise 7: $\angle x =$

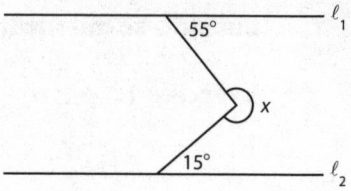

A. 70°

B. 110°

C. 210°

D. 290°

E. 345°

Exercise 8: The ratio of $a°$ to $(a + b)°$ is 3 to 8; $a =$

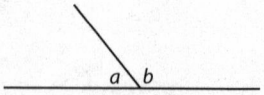

A. 60°

B. 67.5°

C. 72°

D. 108°

E. 112.5°

Now let's look at the answers.

Answer 1: **A:** $3a + a = 180°$; $a = 45°$ and $b = a = 45°$.

Answer 2: **B:** $n = 65°$ and $n + m = 180°$; so $m = 115°$, and $m - n = 50°$.

Answer 3: **B:** $\frac{x}{4} + y + z = 180°$, so $y + z = 180° - \frac{x}{4}$.

Answer 4: **A:** Below the line, $x + 2w = x + w + w = 180°$, so $x + w = 180° - w$.

Answer 5: **A:** This is a toughie. Don't look at vertical angles, look at the supplementary angles. On the bottom, we have $5a + 2a + 5° = 180°$, so $7a = 175°$, and $a = 25°$. Then, on the left, $10b + 5a = 180°$. Substituting $a = 25°$, we get $10b = 180° - 125° = 55°$, or $b = 5.5°$.

If we had looked at the vertical angles after we determined that $a = 25°$, then $10b = 2a + 5° = 2(25°) + 5° = 55°$, so $b = 5.5°$.

Answer 6: **C:** $x + y + 40° = 180°$; so $y = 140° - x$.

Answer 7: **D:** Draw $\ell_3 \parallel \ell_1$ and ℓ_2.

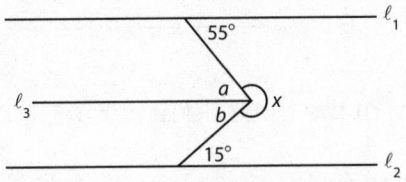

Then $\angle b = 15°$ and $\angle a = 55°$, so

$\angle x = 360° - (15° + 55°) = 290°$.

Answer 8: **B:** One way to answer this exercise is to say $a = \left(\frac{3}{8}\right) \times 180$. You will notice that if you divide 8 into 180, you will have a fraction (or a decimal). So only answer choices B or E could be correct. Because B is < 90, B must be the correct answer. Or if you recognize that $(a + b)° = 180°$, you could use the ratio $\frac{a}{180} = \frac{3}{8}$, which gives 67.5°.

TRIANGLES

Basics about Triangles

A **triangle** is a polygon with three sides. Angles are usually indicated with capital letters. The side opposite the angle is indicated with the same letter, only lowercase.

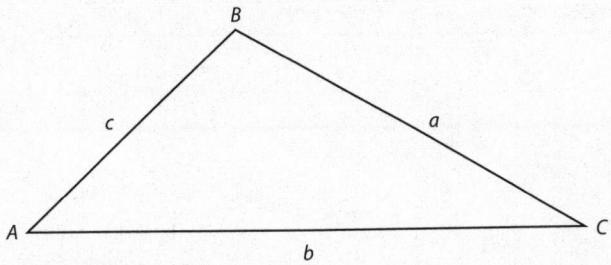

You should know the following general facts about triangles.

- The **sum of the angles** of a triangle is 180°.
- The **altitude**, or **height** (h), of ΔABC shown below is the line segment drawn from a vertex perpendicular to the base, extended if necessary. The **base** of each triangle is $AC = b$.

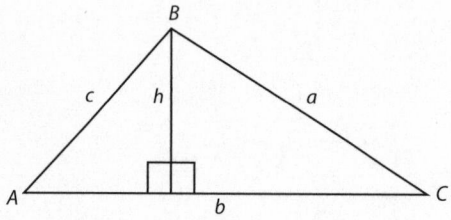

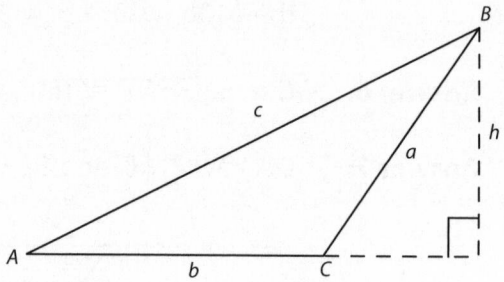

- The **perimeter of a triangle** is the sum of the three sides: $p = a + b + c$.
- The **area of a triangle** is $A = \frac{1}{2}bh$. The reason is that a triangle is half a rectangle. Because the area of a rectangle is base times height, the area of a triangle is half that of a rectangle, as shown in the figure below.

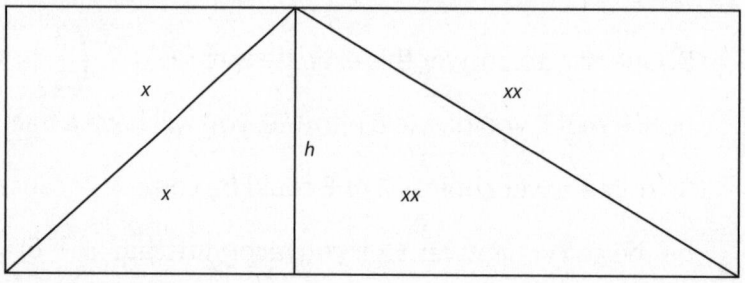

- An **angle bisector** is a line that bisects an angle in a triangle. In the figure below, BD bisects $\angle ABC$ if $\angle 1 = \angle 2$.

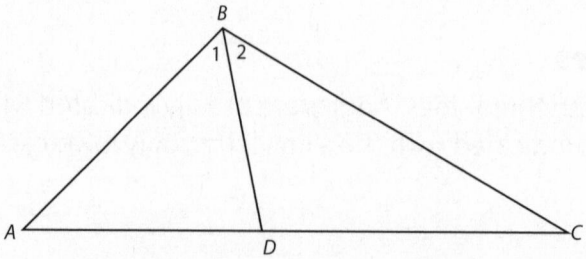

- A **median** is a line drawn from any angle of a triangle to the midpoint of the opposite side. In the figure below, *BD* is a median to side *AC* if *D* is the midpoint of *AC*.

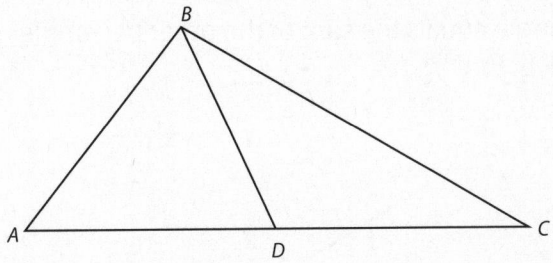

There are many kinds of triangles. One way to describe them is by their sides.

- A **scalene** triangle has three unequal sides and three unequal angles.
- An **isosceles** triangle has at least two equal sides. In the figure below, side *BC* (or *a*) is called the **base**; it may be equal to, greater than, or less than any other side. The **legs**, $AB = AC$ (or $b = c$) are equal. $\angle A$ is the **vertex angle**; it may equal the others, or be greater than or less than the others. The **base angles** are always equal: $\angle B = \angle C$.

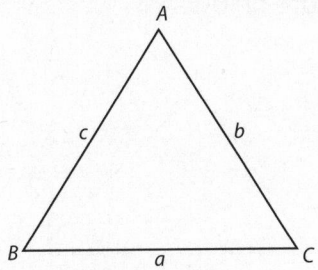

- An **equilateral** triangle is a triangle with all equal sides. All angles equal 60°, so this triangle is sometimes called an **equiangular** triangle. For an equilateral triangle of side *s*, the perimeter $p = 3s$, and the area $A = \frac{s^2\sqrt{3}}{4}$.

Triangles can also be described by their angles.

- An **acute** triangle has three angles that are less than 90°.
- A **right** triangle has one right angle, as shown in the figure below. The **right angle** is usually denoted by the capital letter *C*. The **hypotenuse** *AB* is the side opposite the right angle. The **legs**, *AC* and *BC*, are not necessarily equal. $\angle A$ and $\angle B$ are always **acute** angles.

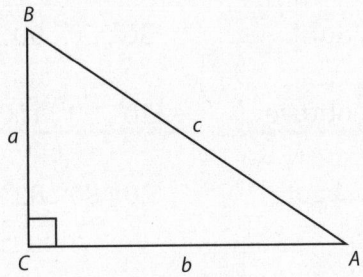

- An **obtuse** triangle has one angle between 90° and 180°.

An **exterior angle** of a triangle is formed by extending one side. In the figure below, $\angle 1$ is an exterior angle. An exterior angle equals the sum of the other two angles (called its two remote interior angles): $\angle 1 = \angle A + \angle B$.

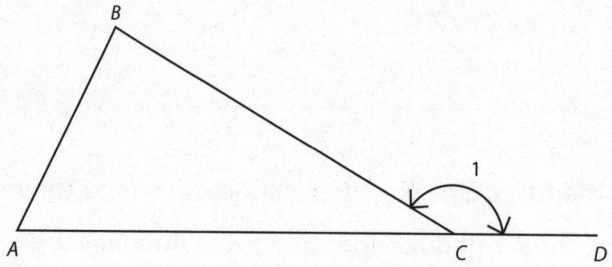

There are three other facts about triangles you should know:

1. The sum of any two sides of a triangle must be greater than the third side.
2. The largest side lies opposite the largest angle, and the largest angle lies opposite the largest side, as shown in the figure below.

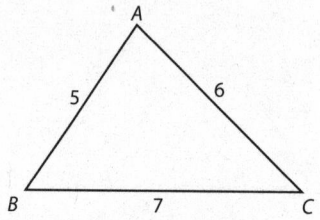

$\angle C < \angle B < \angle A$

3. A line segment joining the midpoints of two sides of a triangle is parallel to the third side and equals half of the third side.

Example 1: Give one set of angles for a triangle that satisfies the following descriptions:

Description	Solution
a. Scalene, acute	50°, 60°, 70°
b. Scalene, right	30°, 60°, 90°. We will deal with this one soon.
c. Scalene, obtuse	30°, 50°, 100°
d. Isosceles, acute	20°, 80°, 80°

e. Isosceles, right	Only one answer: 45°, 45°, 90°. We will deal with this one soon also.
f. Isosceles, obtuse	20°, 20°, 140°
g. Equilateral	Only one answer: three 60° angles

Note *The solutions to Example 1 are not necessarily the only possible answers.*

Let's first do some exercises with angles. Then we'll turn to area and perimeter exercises. We'll finish the chapter with our famous friend Pythagoras and his well-known theorem.

Q **Let's do some more multiple-choice exercises.**

Exercise 9: Two sides of a triangle are 4 and 7. If only integer measures are allowed for the sides, the third side must be taken from which set?

A. {5, 6, 7, 8, 9, 10, 11}
B. {4, 5, 6, 7, 8, 9, 10}
C. {3, 4, 5, 6, 7, 8, 9, 10}
D. {3, 4, 5, 6, 7, 8, 9, 10, 11}
E. {1, 2, 3, 4, 5, 6, 7, 8, 9, 10, 11}

Exercise 10: Arrange the sides in order, largest to smallest, for the figure shown at right.

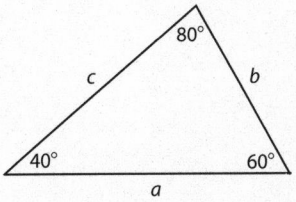

A. *a, b, c*
B. *a, c, b*
C. *b, a, c*
D. *b, c, a*
E. *c, a, b*

Exercise 11: $\angle TVW = 10x$; x could be

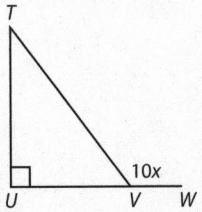

A. 3°
B. 6°
C. 9°
D. 16°
E. 20°

Exercise 12: Write b in terms of a:

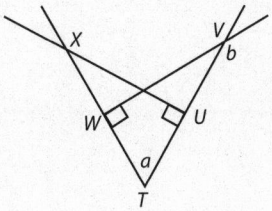

A. $a + 90°$
B. $2a$
C. $2a + 90°$
D. $180° - a$
E. $180° - 2a$

Exercise 13: The largest angle is

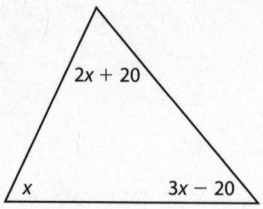

A. 30°

B. 50°

C. 70°

D. 80°

E. 90°

Exercise 14: $\ell \parallel AB$; $y =$

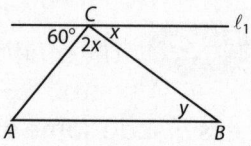

A. 40°

B. 60°

C. 70°

D. 80°

E. Cannot be determined

Use ΔABC for Exercises 15 and 16.

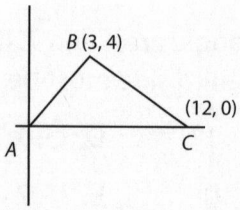

Exercise 15: The area of ΔABC is

A. 18

B. 24

C. 36

D. 48

E. 60

Exercise 16: The perimeter of ΔABC is

A. $17 + \sqrt{97}$

B. 27

C. 32

D. $\sqrt{266}$

E. $10\sqrt{10}$

For Exercises 17 and 18, use this figure of a square with an equilateral triangle on top of it, $AE = 20$.

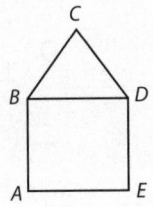

Exercise 17: The perimeter of *ABCDE* is

A. 50

B. 100

C. 120

D. 160

E. 200

Exercise 18: The area of *ABCDE* is

A. 600

B. $100(4 + \sqrt{2})$

C. $100(4 + \sqrt{3})$

D. 800

E. 1,000

For Exercises 19 and 20, use ΔABC with midpoints *X*, *Y*, and *Z*.

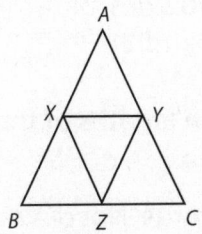

Exercise 19: If the perimeter of ΔABC is 1, the perimeter of ΔXYZ is

A. $\frac{1}{16}$

B. $\frac{1}{8}$

C. $\frac{1}{4}$

D. $\frac{1}{2}$

E. 1

Exercise 20: If the area of ΔABC is 1, the area of ΔXYZ is

A. $\frac{1}{16}$

B. $\frac{1}{8}$

C. $\frac{1}{4}$

D. $\frac{1}{2}$

E. 1

Exercise 21: In the figure shown, $BC = \frac{1}{3}BD$. If the area of $\Delta ABC = 10$, the area of rectangle $ABDE$ is

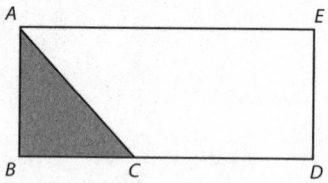

A. 30

B. 40

C. 60

D. 120

E. Cannot be determined

Let's look at the answers.

Answer 9: **B:** The third side s must be greater than the difference and less than the sum of the other two sides, or $> 7 - 4$ and $< 7 + 4$. Thus the third side must be between 3 and 11.

Answer 10: **B:** Judge the relative lengths of the sides by the size of the angles opposite them. Then $a > c > b$.

Watch out for the words "Not drawn to scale." If it is a simple figure, "not drawn to scale" usually means it is not drawn to scale, and you cannot assume relative sizes without being given actual measurements. However, if it is a semi-complicated or complicated figure, the figure probably *is* drawn to scale.

Answer 11: **D:** $\angle TVW$ must be between 90° and 180°, so $9° < x < 18°$.

Answer 12: **A:** This is really tricky. UX is drawn to confuse you. In ΔTVW, b is the exterior angle, so $b = a + 90°$.

Answer 13: **D:** $x + 2x + 20 + 3x - 20 = 180$, or $6x = 180$, so $x = 30$. $2x + 20 = 80$ and $3x - 20 = 70$. The largest angle is 80°.

Answer 14: **A:** $2x + x + 60 = 180$; $x = 40°$. But $y = x = 40°$ (because $\ell_1 \parallel AB$).

Answer 15: **B:** $A = \frac{1}{2}bh = \frac{1}{2} \times 12 \times 4 = 24.$

Answer 16: **A:** Use the distance formula to find sides *AB* and *BC*. $p = AC + AB + BC = 12 + \sqrt{4^2 + 3^2} + \sqrt{(3 - 12)^2 + (4 - 0)^2} = 12 + \sqrt{25} + \sqrt{97} = 17 + \sqrt{97}.$

Answer 17: **B:** Do not include *BD*; $p = 5 \times 20 = 100.$

Answer 18: **C:** Area $= s^2 + \frac{s^2\sqrt{3}}{4} = 20^2 + \frac{20^2\sqrt{3}}{4} = 400 + 100\sqrt{3} = 100\,(4 + \sqrt{3}).$

Answer 19: **D:** All the sides of ΔXYZ are medians of ΔABC. If the perimeter of ΔABC is 1, then all the sides of ΔXYZ are half of those of ΔABC, and so is the perimeter.

Answer 20: **C:** If the sides of ΔXYZ are half of those of ΔABC, the area of ΔXYZ is $\left(\frac{1}{2}\right)^2 A = \frac{1}{4}A = \frac{1}{4}.$

Answer 21: **C:** If we draw lines parallel to *DE* to divide the original rectangle into three congruent rectangles, and then divide each rectangle into two triangles, we see that each triangle is one-sixth of the rectangle. So the area of the rectangle is $6(10) = 60.$

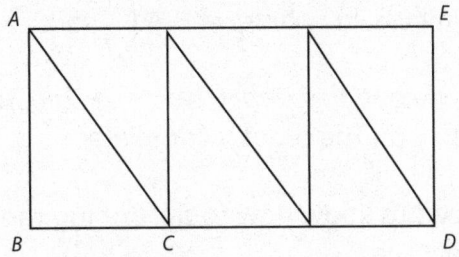

We'll have more of these type of exercises as part of Chapter 12, where we discuss circles.

Let's go on to good old Pythagoras.

PYTHAGOREAN THEOREM

This is perhaps the most famous math theorem of all. Most theorems have one proof. A small fraction have two. This theorem, however, has more than a hundred, including three by past presidents of the United States. We've had some smart presidents who actually knew some math.

The Pythagorean Theorem simply states:

In a right triangle, the hypotenuse squared is equal to the sum of the squares of the legs.

In symbols,

$$c^2 = a^2 + b^2.$$

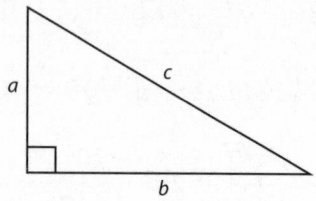

As a teacher, I must show you one proof.

Proof:

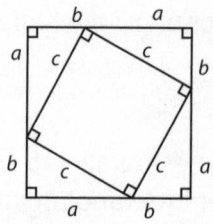

In this figure, the area of the larger square equals that of the smaller square plus the four congruent triangles. In symbols, $(a + b)^2 = c^2 + 4\left(\frac{1}{2}ab\right)$.

Multiplying this equation out, we get $a^2 + 2ab + b^2 = c^2 + 2ab$. Then subtracting $2ab$ from both sides, we get $c^2 = a^2 + b^2$. The proof is complete.

There are two basic problems you need to know how to do: finding the hypotenuse and finding one of the legs of the right triangle.

Example 2: Solve for x:

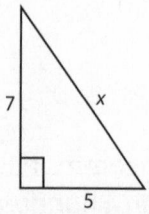

Solution: $x^2 = 7^2 + 5^2; x = \sqrt{74}$.

Example 3: Solve for x:

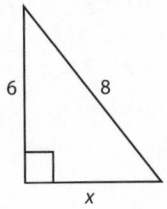

Solution: $8^2 = 6^2 + x^2$, or $x^2 = 64 - 36 = 28$. So $x = \sqrt{28} = \sqrt{2 \times 2 \times 7} = 2\sqrt{7}$.

Notice that the hypotenuse squared is always by itself, whether it is a number or a letter.

Pythagorean Triples

It is a good idea to memorize these triples. Those in the 3-4-5 and 5-12-13 groups are just multiples of the groups. Below are the sides of the most important right triangles. The hypotenuse is always listed third in each group.

The 3-4-5 group:	3-4-5, 6-8-10, 9-12-15, 12-16-20, 15-20-25
The 5-12-13 group:	5-12-13, 10-24-26
The rest:	8-15-17, 7-24-25, 20-21-29, 9-40-41, 11-60-61

Special Right Triangles

You ought to know two other special right triangles, the isosceles right triangle (with angles 45°-45°-90°) and the 30°-60°-90° right triangle. The facts about these triangles can all be found by using the Pythagorean Theorem.

1. The 45°-45°-90° isosceles right triangle:
   - The legs are equal.
   - To find a leg given the hypotenuse, divide the hypotenuse by $\sqrt{2}$ (or multiply by $\frac{\sqrt{2}}{2}$).
   - To find the hypotenuse given a leg, multiply the leg by $\sqrt{2}$.

Example 4: Find x and y for this isosceles right triangle.

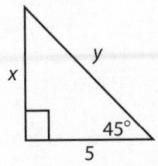

Solution: $x = 5$ (the legs are equal); $y = 5\sqrt{2}$.

Example 5: Find x and y for this isosceles right triangle.

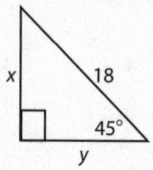

Solution: $x = y = \frac{18}{\sqrt{2}} = 18 \times \frac{\sqrt{2}}{2} = 9\sqrt{2}$.

2. The 30°-60°-90° right triangle.
   - If the shorter leg (opposite the 30° angle) is not given, get it first. It is always half the hypotenuse.
   - To find the short leg given the hypotenuse: divide the hypotenuse by 2.
   - To find the hypotenuse given the short leg: multiply the short leg by 2.
   - To find the short leg given the long leg: divide the long leg by $\sqrt{3}$ (or multiply by $\frac{\sqrt{3}}{3}$).
   - To find the long leg given the short leg: multiply the short leg by $\sqrt{3}$.

Example 6: Find x and y for this right triangle.

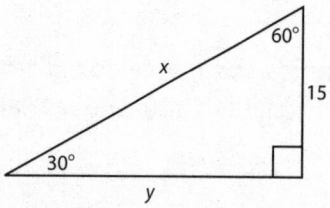

Solution: The short leg is given (15); $x = 2(15) = 30$; $y = 15\sqrt{3}$.

Example 7: Find x and y for this right triangle.

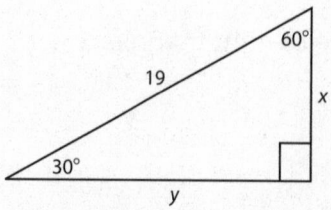

Solution: $x = \frac{19}{2} = 9.5$; $y = 9.5\sqrt{3}$.

Example 8: Find x and y for this right triangle.

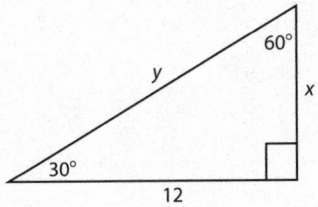

Solution: $x = \frac{12}{\sqrt{3}} = 12\frac{\sqrt{3}}{3} = 4\sqrt{3}$; $y = 2(4\sqrt{3}) = 8\sqrt{3}$.

Q Let's do a few more multiple-choice exercises.

Exercise 22: Two sides of a right triangle are 3 and $\sqrt{5}$.

I: The third side is 2.

II: The third side is 4.

III: The third side is $\sqrt{14}$.

Which of the following choices is correct?

A. Only II is true.

B. Only III is true.

C. Only I and II are true.

D. Only I and III are true.

E. I, II, and III are true.

Exercise 23: The area of square $ABCD$ =

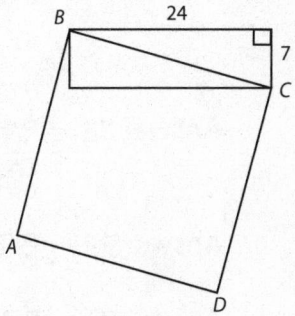

A. 50

B. 100

C. 225

D. 576

E. 625

Exercise 24: x =

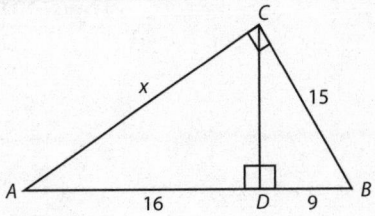

A. 16

B. 18

C. 20

D. 22

E. 24

Exercise 25: $c^2 - b^2 =$

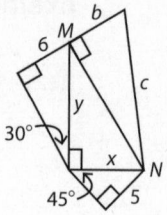

A. 72

B. 144

C. 216

D. 194

E. 288

Exercise 26: $x =$

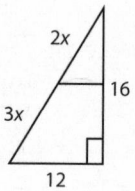

A. 1

B. 2

C. 3

D. 4

E. 4.5

Exercise 27: A 25-foot ladder is leaning on the wall. Its base is 15 feet from the wall. If the ladder is pushed until it is only 7 feet from the wall, how much farther up the wall is the ladder pushed?

A. 4 feet

B. 8 feet

C. 12 feet

D. 20 feet

E. 24 feet

Let's look at the answers.

Answer 22: **D:** Try the Pythagorean Theorem with various combinations of 3, $\sqrt{5}$, and x (the third side). The only ones that work are Statement I: $2^2 + \left(\sqrt{5}\right)^2 = 3^2$; and Statement III: $3^2 + \left(\sqrt{5}\right)^2 = \left(\sqrt{14}\right)^2$.

Answer 23: **E:** We recognize the right triangle as a 7-24-25 triple, so side $BC = 25$. The area of the square is $(25)^2 = 625$.

Answer 24: **C:** This is a 15-20-25 triple, so $x = 20$.

Answer 25: **D:** We see that MN is the side of two triangles. By the Pythagorean Theorem, we get $c^2 - b^2 = x^2 + y^2 = \left(5\sqrt{2}\right)^2 + 12^2 = 50 + 144 = 194$.

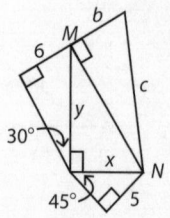

Answer 26: **D:** This triangle is a 12-16-20 triple, so $3x + 2x = 5x = 20$, and $x = 4$.

Answer 27: **A:** The first figure is a 15-20-25 right triangle with the ladder 20 feet up the wall. The second figure is a 7-24-25 triple with the ladder 24 feet up the wall. The ladder is pushed another $24 - 20 = 4$ feet up the wall.

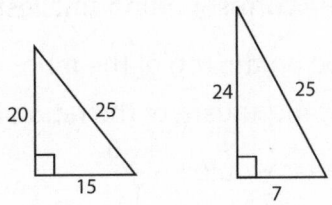

SIMILAR TRIANGLES

Two triangles are similar if the corresponding angles are congruent and corresponding sides are in proportion.

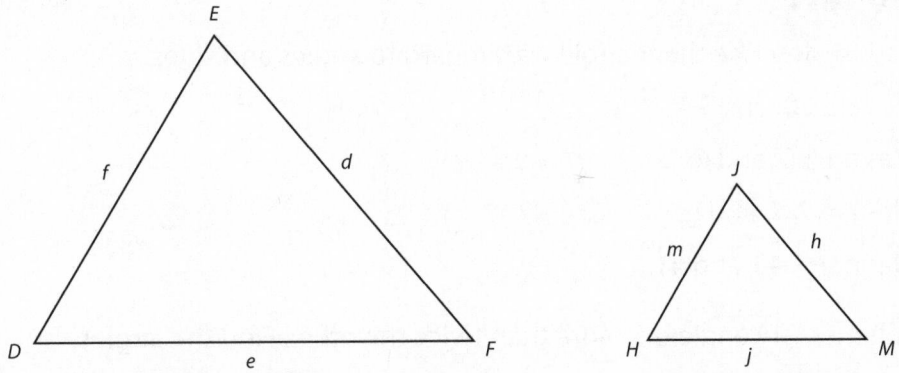

In the above figure, this means $\angle D \cong \angle H$, $\angle E \cong \angle J$, and $\angle F \cong \angle M$ and $\frac{f}{m} = \frac{d}{h} = \frac{e}{j}$.

$\frac{f}{m} = \frac{d}{h} = \frac{e}{j}$ is called the **ratio of similarity**.

Example 9: For the figures below,

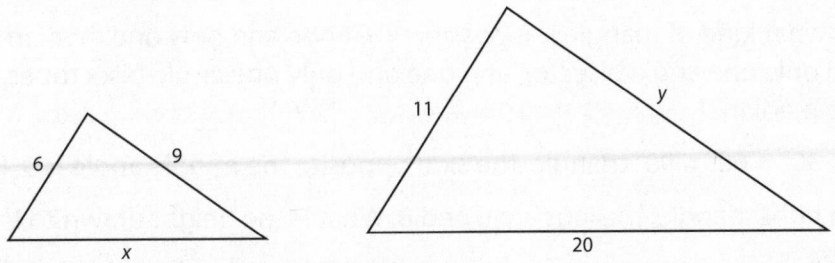

a. What is the ratio of similarity?

b. Find x and y.

c. What is the ratio of their areas?

Solutions:

a. The left sides in each triangle are given. The ratio of similarity is $\frac{6}{11}$.

b. $\frac{6}{11} = \frac{x}{20}$; $11x = 120$; $x = \frac{120}{11}$; $\frac{6}{11} = \frac{9}{y}$; $6y = 99$; $y = \frac{99}{6}$ or $\frac{33}{2}$. Once you start with the length of the smallest side of the first triangle on the top of a fraction, the corresponding smallest side of the second triangle must always be on the top of the ratio.

c. The ratio of the areas is the square of the ratio of similarity. In this case, the ratio of the areas would be $\frac{36}{121}$.

Note *For three-dimensional similar figures, the ratio of their volumes would be the cube of the ratio of similarity.*

Chapter 10 Test

For questions 1–4, describe the triangle with regard to angles and sides.

1. Two angles are 20° and 30°.
2. Two angles are 60° and 60°.
3. Three sides are 7, 7, and 6.
4. Three sides are 9, 40, and 41.
5. One supplementary angle is 3 more than twice the other. Find the angles.
6. The base of an isosceles triangle is 6 more than a leg. The perimeter is 99 feet. Find the sides.
7. The hypotenuse is 8 and one leg is 7; find the area and perimeter of the triangle.
8. The area of an equilateral triangle is $100\sqrt{3}$. Find its perimeter.
9. The ratio of similarity of two triangles is 2 to 3. If 24 is the area of the smaller triangle, find the area of the larger triangle.
10. For what kind of triangle is it possible that one and only one median equals one and only one angle bisector, and one and only one angle bisector equals one and only one median?
11. In a 30° = 60° = 90° triangle, the side opposite the middle angle is 9. Find its area.
12. In a right triangle, the legs are 6 and 8. What is the height drawn to the hypotenuse?
13. A regular hexagon consists of six equilateral triangles. One side is 4; find the perimeter and area.
14. An isosceles triangle has legs of 15. If the base is 24, find its area.
15. If two vertical angles are $7x - 3$ and $2x + 17$, what are their measures?

16. Two interior angles on the same side of a transversal cutting two parallel lines are $3x + 5$ and $5x + 15$. Find the angles.
17. An isosceles right triangle has an area of 50. Find the hypotenuse.
18. Find the side of an equilateral triangle if a square of side 25 and the triangle have the same perimeter.
19. The second side of a triangle is 3 less than the first, and the third side is 9 less than the second. If the perimeter is 90, find the sides.
20. If the area of a square is equal to the area of an isosceles right triangle, write the leg of the right triangle in terms of s, the side of the square.

Answers to Chapter 10 Test

1. The third angle is $180° - 50° = 130°$; obtuse and scalene.
2. The third angle is also 60°; equilateral and equiangular (acute also OK).
3. Isosceles and acute.
4. Right and scalene.
5. $x + 2x + 3 = 180$; $x = 59°$; $2x + 3 = 121°$.
6. $x = \text{leg}$; $x + x + x + 6 = 99$; $x = 31$; the three sides are 31, 31, and 37.
7. The other leg $= \sqrt{8^2 - 7^2} = \sqrt{15}$; $A = \frac{1}{2}(\ell_1 \times \ell_2) = \frac{7\sqrt{15}}{2}$; the perimeter is $15 + \sqrt{15}$.
8. $A = \frac{s^2\sqrt{3}}{4} = 100\sqrt{3}$. So $s = 20$ and $p = 60$.
9. The ratio of the areas is $\frac{4}{9}$; $\frac{4}{9} = \frac{24}{x}$; $x = 54$.
10. Isosceles triangle from the vertex to the base or an equilateral triangle.
11. The short leg is $\frac{9}{\sqrt{3}} = 3\sqrt{3}$; $A = \frac{1}{2}b \times h = \frac{1}{2}9(3\sqrt{3}) = 13.5\sqrt{3}$.
12. The hypotenuse is 10; $A = \frac{1}{2}b_1h_1 = \frac{1}{2}b_2h_2$; substitute the values and multiply both sides by 2 to get $(6)(8) = 10(h_2)$; $h_2 = 4.8$.
13. $p = 6(4) = 24$; $A = 6 \times \frac{s^2\sqrt{3}}{4} = 24\sqrt{3}$.
14. The height from the vertex bisects the base; we have a 9 – 12 – 15 triangle; the height is 9; $A = \frac{1}{2}(bh) = \frac{1}{2}(24)(9) = 108$.
15. $7x - 3 = 2x + 17$; $x = 4°$; the vertical angles are 25° each.
16. $3x + 5 + 5x + 15 = 180$; $x = 20°$; the angles are 65° and 115°.

17. $A = \frac{1}{2}\ell^2 = 50; \ell^2 = 100; \ell = 10; h = \ell\sqrt{2} = 10\sqrt{2}.$

18. $4(25) = 3s; s = 33\frac{1}{3}.$

19. $x + (x - 3) + (x - 12) = 90; 3x - 15 = 90; x = 35$; the answers are 35, 32, and 23.

20. If the areas are the same, $s^2 = \frac{1}{2}\ell^2$, and $\ell = s\sqrt{2}$.

That's all for angles and triangles for now. We will see more when circles are discussed in Chapter 12. For now, though, let's look at rectangles and other polygons.

Answer to "Bob Asks": Sequoia

CHAPTER 11: *Other Two-Dimensional Figures*

"Sue had a 20% salary increase, followed by a 20% salary decrease. Lou had a 20% salary decrease, followed by a 20% salary increase. Who got the better deal? What was the difference increase or decrease between Sue and Lou, or was it the same?"

We now deal with the rest of the polygons (closed figures with line-segment sides).

QUADRILATERALS

Parallelograms

A **parallelogram** is a **quadrilateral** (four-sided polygon) with parallel opposite sides.

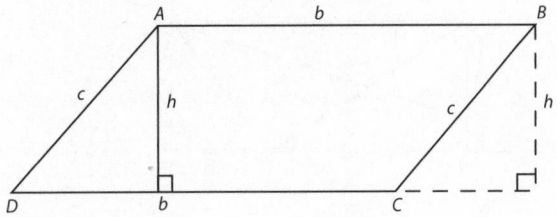

You should know the following properties about parallelograms:

- The opposite angles are equal. $\angle DAB = \angle BCD$ and $\angle ADC = \angle ABC$.
- The consecutive angles are supplementary. $\angle DAB + \angle ABC = \angle ABC + \angle BCD = \angle BCD + \angle CDA = \angle CDA + \angle DAB = 180°$.
- The opposite sides are equal. $AB = CD$ and $AD = BC$.
- Area $= A = bh$.
- Perimeter $= p = 2b + 2c$.
- The diagonals bisect each other. $AE = EC$ and $DE = EB$ in the figure below.

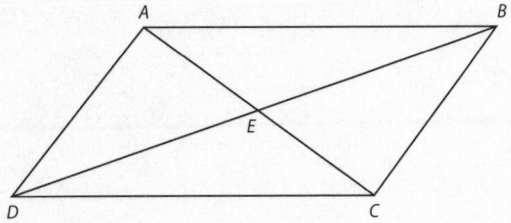

Example 1: For parallelogram *RSTU* shown in the figure below, find the following if $RU = 10$:

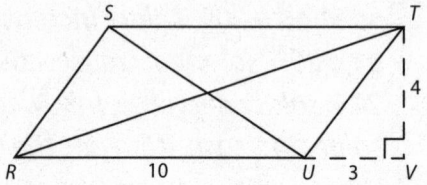

a. The area

b. The perimeter

c. Diagonal *RT*

d. Diagonal *SU*

Solutions:

a. $A = bh = (10)(4) = 40$ square units. The whole test should be this easy!

b. We have to find the length of $RS = TU$. $TU = 5$ because it is the hypotenuse of a 3-4-5 right triangle. So the perimeter is $p = 2(10) + 2(5) = 30$ units.

c. $RT = \sqrt{(RV)^2 + (TV)^2} = \sqrt{13^2 + 4^2} = \sqrt{185}$

d.

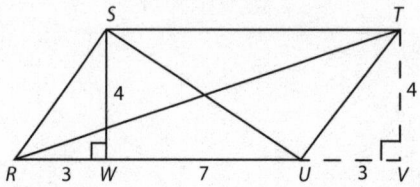

To find diagonal *SU*, draw the other altitude *SW* as pictured.

$SU = \sqrt{WU^2 + SW^2} = \sqrt{7^2 + 4^2} = \sqrt{65}$

Example 2: For parallelogram *WXYZ* with altitudes *XM* and *YN*, find the following in terms of *a*, *b*, and *c*:

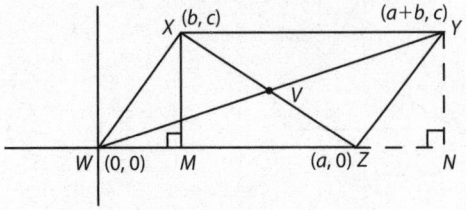

a. The coordinates of point *M*

b. The coordinates of point *N*

c. The coordinates of point *V*

d. The perimeter

e. The area

Solutions:

a. *M* has the same *x*-coordinate as point *X* and the same *y*-coordinate as point *W*, so the coordinates of *M* are $(b, 0)$.

b. *N* has the same *x*-coordinate as point *Y* and the same *y*-coordinate as point *W*, so the coordinates of *N* are $(a + b, 0)$.

c. *V* is halfway between *W* and *Y*, so use the formula for the midpoint between $Y(a + b, c)$ and $W(0, 0)$: Midpoint $V = \left(\frac{(a+b)+0}{2}, \frac{c+0}{2}\right) = \left(\frac{a+b}{2}, \frac{c}{2}\right)$.

d. $WZ = XY$ is length *a*. By the distance formula, $WX = ZY = \sqrt{(b-0)^2 + (c-0)^2} = \sqrt{b^2 + c^2}$. Therefore, the perimeter is $p = 2a + 2\sqrt{b^2 + c^2}$.

e. Area $= A =$ base $\times$ height $= ac$.

Example 3: For parallelogram *EFGH*, find the smaller angle.

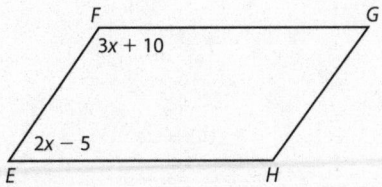

Solution: Consecutive angles of a parallelogram are supplementary. Therefore, $(3x+10)° + (2x - 5)° = 180°$; $x = 35°$; so the smaller angle is $2(35°) - 5° = 65°$. Be careful to give the answer the ACCUPLACER wants. Two other, but incorrect, answer choices would be 35° and 115°, for those who do not read carefully!!!

Rhombus

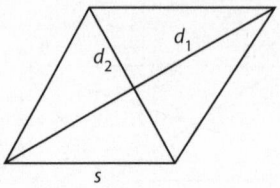

A **rhombus** is an equilateral parallelogram.

Thus, a rhombus has all of the properties of a parallelogram plus the following:

- All sides are equal.
- The diagonals are perpendicular bisectors of each other.
- Perimeter $= p = 4s$.
- Area $= A = bh = \frac{1}{2} \times d_1 \times d_2$, the area equals half the product of its diagonals.

Example 4: For the rhombus given below with side $s = 13$ and larger diagonal $BD = 24$, find the other diagonal and the area.

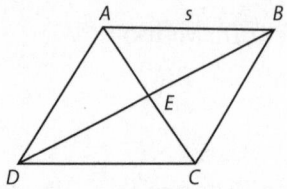

Solution: $AB = 13$ and $BD = 24$. Because the diagonals bisect each other, $BE = 12$. The diagonals are perpendicular to each other, so ΔABE is a 5-12-13 right triangle, and $AE = 5$. Therefore, the other diagonal $AC = 10$. The area is $A = \frac{1}{2} \times BD \times AC = \frac{1}{2}(24)(10) = 120$ square units.

Example 5: Find the area of a rhombus with side 10 and smaller interior angle of 60°.

Solution: If we draw the diagonal through the two larger angles, we will have two congruent equilateral triangles. The area of this rhombus is twice the area of each triangle, or $2 \times \frac{s^2\sqrt{3}}{4}$. Because $s = 10$, the area is $A = 2 \times \frac{10^2\sqrt{3}}{4} = 50\sqrt{3}$ square units.

Now let's go on to more familiar territory.

Rectangle

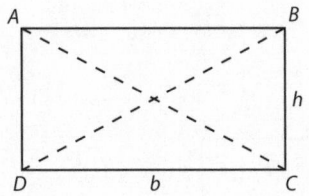

A **rectangle** is a parallelogram with right angles.

Therefore, it has all of the properties of a parallelogram plus the following:

- All angles are 90°.
- Diagonals are equal (but *not* perpendicular).
- Perimeter $= p = 2b + 2h$.
- Area $= A = bh$. This is a **postulate** (law taken to be true without proof) from which we get the area of all other figures with sides that are line segments.

Example 6: One base of a rectangle is 8, and one diagonal is 9. Find all the sides and the other diagonal. Find the perimeter and area.

Solution: The top base and bottom base are both 8. Both diagonals are 9. The other two sides are each $\sqrt{9^2 - 8^2} = \sqrt{17}$. So the perimeter is $p = 16 + 2\sqrt{17}$ units; and the area is $A = 8\sqrt{17}$ square units.

Example 7: $AB = 10$, $BC = 8$, $EF = 6$, and $FG = 3$. Find the area of the shaded region of the figure below.

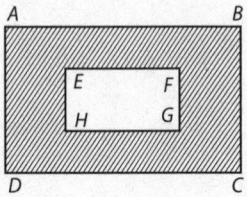

Solution: The area of the shaded region is the area of the outside rectangle minus the area of the inside one. $A = (10)(8) - (6)(3) = 62$ square units.

Example 8: In polygon $ABCDEF$, $BC = 30$, $AF = 18$, $AB = 20$, and $CD = 11$. Find the perimeter and area of the polygon.

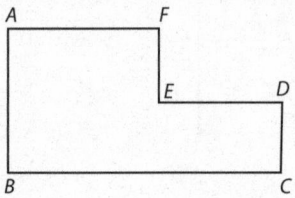

Solution: Draw a line through DE, hitting AB at point G. Then $AF = GE$ and $BC = GD$. Because $DG = 30$ and $GE = 18$, $DE = 12$. $AB = CD + EF$. $AB = 20$ and $CD = 11$, so $EF = 9$. This gives the lengths of all the sides. The perimeter thus is $p = AB + BC + CD + DE + EF + AF = 20 + 30 + 11 + 12 + 9 + 18 = 100$ units.

The area of rectangle $BCDG$ is $BC \times CD = (30)(11) = 330$. The area of rectangle $AFEG$ is $AF \times FE = (18)(9) = 162$. Therefore, the total area is $330 + 162 = 492$ square units. There are other ways to find this area, as you might be able to see.

Example 9: The areas of the pictured rectangle and triangle are the same. If $\frac{LW}{4} = 20$, what is bh?

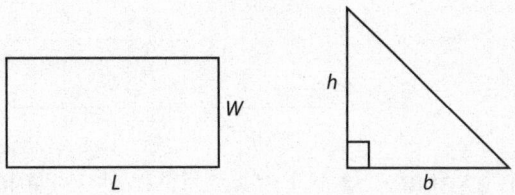

Solution: $\frac{LW}{4} = 20$; so $LW = 80$, which is the area of the rectangle. Because that also is the area of the triangle, $\frac{1}{2}bh = 80$. So $bh = 160$.

Square

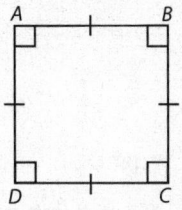

A **square** is a rectangle with equal sides, or it can be thought of as a rhombus with four equal 90° angles.

Therefore, it has all of the properties of a rectangle and a rhombus:

- All sides are equal.
- All angles are 90°.
- Both diagonals bisect each other, are perpendicular to each other, and are equal.
- Each diagonal $d = d_1 = d_2 = s\sqrt{2}$, where s = a side.
- Perimeter $= p = 4s$.
- Area $= A = \frac{d^2}{2} = s^2$.

Example 10: What is the area of this square?

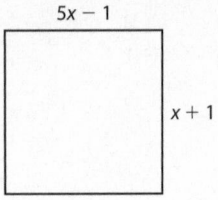

Solution: Because it is a square, $5x - 1 = x + 1$, so $x = \frac{1}{2}$ and $x + 1 = 1\frac{1}{2}$. Then $A = \left(1\frac{1}{2}\right)^2 = 2\frac{1}{4}$.

Example 11: The area of square *C* is 36; the area of square *B* is 25. What is the area of square *A*?

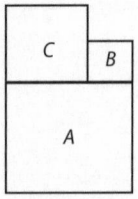

Solution: The side of square *C* must be 6, and the side of square *B* must be 5. Therefore, the side of square *A* is 11, and the area of square *A* is $11^2 = 121$.

Use the figure below for Examples 12 and 13. It is a square surmounted by an equilateral triangle (I've always wanted to write that word). $AB = 10$.

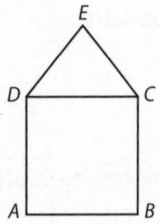

Example 12: What is the perimeter of the figure?

Solution: The perimeter is $5(10) = 50$. Note that *CD* is *not* part of the perimeter.

Example 13: What is the area of the figure?

Solution: $A = s^2 + \frac{s^2\sqrt{3}}{4} = 10^2 + \frac{10^2\sqrt{3}}{4} = 100 + 25\sqrt{3} = 25(4 + \sqrt{3})$.

Trapezoid

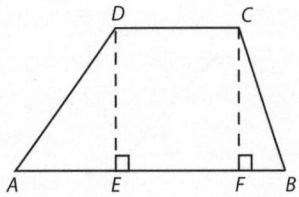

A **trapezoid** is a quadrilateral with exactly one pair of parallel sides.

Because a trapezoid is *not* a type of parallelogram, it has its own unique set of properties, as follows:

- The parallel sides, *AB* and *CD*, are called **bases**.
- The heights, *DE* and *CF*, are equal.
- The legs, *AD* and *BC*, may or may not be equal.
- The diagonals, *AC* and *BD*, may or may not be equal.
- Perimeter $= p = AB + BC + CD + DA$.
- Area $= A = \frac{1}{2}h(b_1 + b_2)$, where b_1 and b_2 are the bases and h is the height.

Note *If we draw one of the diagonals, we see that a trapezoid is the sum of two triangles. Factoring out* $\frac{1}{2}h$, *we get the formula for the area of the trapezoid.*

If the legs are equal, the trapezoid is called an **isosceles trapezoid**, shown below.

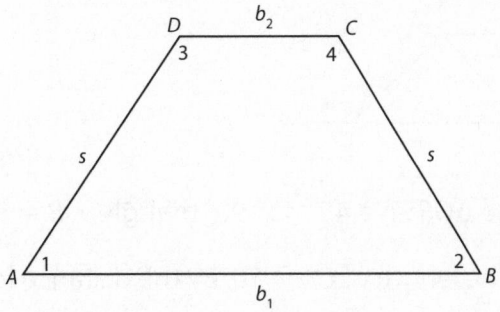

An isosceles trapezoid has the following additional properties:

- Perimeter $= p = b_1 + b_2 + 2s$.
- The diagonals are equal, $AC = BD$.
- The base angles are equal, $\angle 1 = \angle 2$ and $\angle 3 = \angle 4$.

Example 14: Find the area and the perimeter of Figure *ABCD*.

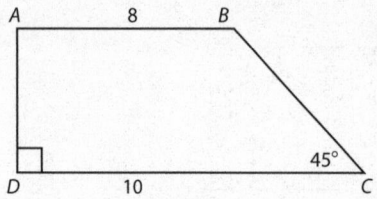

Solution: Draw the height from point *B*, *BG*, as shown below.

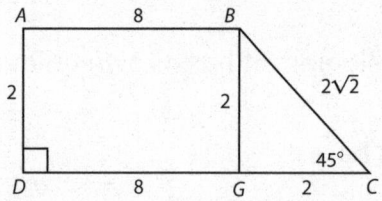

$DG = 8$; $CG = 2$. Because ΔBGC is an isosceles right triangle, the height BG (and AD) $= 2$; BC, the hypotenuse of the isosceles right triangle, is therefore $2\sqrt{2}$. Therefore, the perimeter is $p = 10 + 2 + 8 + 2\sqrt{2} = 20 + 2\sqrt{2}$, and the area is $A = \frac{1}{2}h(b_1 + b_2) = \frac{1}{2}(2)(8 + 10) = 18$.

Example 15: Given trapezoid *ORST*, with $RS \parallel OT$, find the coordinates of point *S*. Find the perimeter and the area of trapezoid *ORST*.

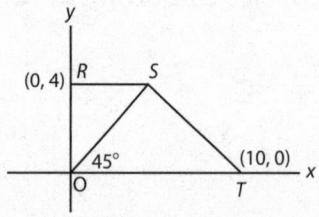

Solution: Because ΔORS is a 45°-45°-90° triangle, $OR = RS = 4$, so *S* is the point (4, 4). The length of $OT = 10$. By the distance formula, the length of $ST = \sqrt{(10 - 4)^2 + (0 - 4)^2} = \sqrt{52} = 2\sqrt{13}$. Therefore, the perimeter is $p = 10 + 4 + 4 + 2\sqrt{13}$, or, to be fancy, $2(9 + \sqrt{13})$. Area $= \frac{1}{2}h(b_1 + b_2)$ $= \frac{1}{2}(4)(10 + 4) = 28$.

Example 16: Find the area of isosceles trapezoid *EFGH*.

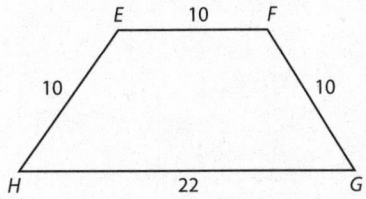

Solution: Draw in the two heights for the trapezoid.

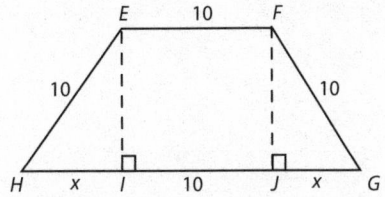

The two bases of the triangles formed are equal because it is an isosceles trapezoid. From the figure, $2x + 10 = 22$, so $x = 6$. Each of the triangles is a 6-8-10 Pythagorean triple, so the height of the trapezoid is 8. Therefore, $A = \frac{1}{2}(8)(10 + 22) = 128$.

POLYGONS

Let's talk about polygons in general. Most of the time, we deal with **regular** polygons. A regular polygon has all sides equal and all angles equal. A square and an equilateral triangle are examples of regular polygons we have already discussed.

Any *n*-sided polygon has the following properties.

- The sum of all the interior angles is $(n - 2)180°$, where n is the number of sides (or angles) in the polygon.
- The sum of all exterior angles of a polygon always equals 360°, no matter how many sides it has.
- An interior angle plus its exterior angle always add to 180°.
- The number of diagonals is $\frac{n(n-3)}{2}$.

If the polygon is regular, it has the following additional properties:

- Each exterior angle $= \frac{360°}{n}$.
- Each interior angle $= \frac{(n-2)180°}{n}$.

Polygons are named for the number of sides they have.

A **pentagon** is a 5-sided polygon.

A **hexagon** is a 6-sided polygon.

A **heptagon** is a 7-sided polygon.

An **octagon** is an 8-sided polygon.

A **nonagon** is a 9-sided polygon.

A **decagon** is a 10-sided polygon.

A **dodecagon** is a 12-sided polygon.

An **n-gon** is an *n*-sided polygon.

Example 17: An octagon has a perimeter of 27. If 5 is added to each side, what is the perimeter of the new octagon?

Solution: It doesn't matter how long each side is! If 5 is added to each of 8 sides, 40 is added to the perimeter. The new perimeter is $27 + 40 = 67$.

Example 18: The side of a regular hexagon is 4. Find its area.

Solution: A regular hexagon is made up of six equilateral triangles. The side of each triangle is 4, and the area is $A = 6\frac{s^2\sqrt{3}}{4} = 6\frac{4^2\sqrt{3}}{4} = 24\sqrt{3}$.

Example 19: The sum of the interior angles of a polygon is 720°. Find the number of sides, the number of degrees in one exterior angle, and the number of degrees in one interior angle.

Solution: $(n - 2)(180) = 720$. Divide each side of this equation by 180 to simplify: $n - 2 = 4$, so $n = 6$ sides. One exterior angle $= \frac{360°}{6} = 60°$. An interior angle is supplemental to its external angle, by definition, so $180° - 60° = 120°$ for each interior angle.

Chapter 11 Test

For questions 1–4, identify the triangle with regard to angles and sides.

1. In a parallelogram, one angle is $3x - 7$. The opposite angle is $x + 37$. Find the four angles.
2. The bases of a trapezoid are 10 and 8. If the height is 7, find the area.
3. The side of a square is 10. Find the base of a rectangle of the same area if the height is 12.
4. How many diagonals are in a pentagon?
5. How many degrees are in the sum of the interior angles of a hexagon?
6. The angles of a quadrilateral are in the ratio of 4 to 5 to 5 to 6. Find the angles.

For questions 7–9, *ABCD* is a quadrilateral with $A(1, 4)$, $B(10, 9)$, $C(14, 15)$, $D(5, 10)$.

7. Show that quadrilateral *ABCD* is a parallelogram from the definition.
8. Show that the opposite sides of this quadrilateral are equal.
9. Show that the diagonals of this parallelogram bisect each other.
10. A parallelogram has base 20, slanted side 10, and the angle between them is 30°. Find its area.
11. The area of a square is 200. Find its diagonal.
12. If you double the base and triple the height, what happens to the area of a rectangle?
13. A rectangle is 5 yards by 2 yards. If it cost $4.00 a square foot to carpet the rectangle, how much will you pay?
14. A rhombus has diagonals 12 and 16. Find its area.
15. The interior angle of a regular polygon is 144°. What is its name?
16. The diagonal of a rectangle is 11, and one side is 9. Find its area.
17. For question 16, find the perimeter of the rectangle.
18. The perimeter and area of a square are the same number. What is the length of its side?
19. If a trapezoid and a triangle have the same base and same height, why is the area of the trapezoid always larger?
20. The perimeter of a rectangle is 24 and the area is 20. Find the length and width.

Answers to Chapter 11 Test

1. $3x - 7 = x + 37$; $x = 22$; the angles are 59°, 59°, 121°, and 121°.
2. $A = \frac{1}{2}h(b_1 + b_2) = \frac{7}{2}(10 + 8) = 63.$

3. $s^2 = bh$; $10^2 = 12b$; $b = 8.3$.
4. $\frac{n(n-3)}{2} = \frac{5(5-3)}{2} = 5$.
5. $(n - 2)\,180 = (6 - 2)\,180° = 720°$.
6. $4x + 5x + 5x + 6x = 360°$; 72°, 90°, 90°, 108°.
7. Opposite sides must be parallel; the slopes of *AB* and *CD* are the same $\left(\frac{5}{9}\right)$, and the slopes of *AD* and *BC* are the same $\left(\frac{3}{2}\right)$.
8. $AB = CD = \sqrt{106}$, and $AD = BC = 2\sqrt{13}$.
9. The midpoint of *AC* and *BD* is (7.5, 9.5); since they are both the same point, they bisect each other.
10. We form a 30°-60°-90° right triangle; the side opposite the 30° angle is half the hypotenuse (which is 10); so the height is 5, and the area is 100.
11. $s^2 = 200$; $s = 10\sqrt{2}$; so $d = s\sqrt{2} = 10\sqrt{2} \times \sqrt{2} = 20$.
12. Multiplied by 6.
13. 15 by 6, or 90 square feet times \$4.00 = \$360.
14. $A = \frac{1}{2} d_1 \times d_2 = 96$.
15. Exterior angle = 180° − 144 = 36°; $\frac{360}{36} = 10$ sides; decagon.
16. Other side $= \sqrt{11^2 - 9^2} = 2\sqrt{10}$. $A = 9 \times 2\sqrt{10} = 18\sqrt{10}$.
17. $P = 18 + 4\sqrt{10}$ or $2(9 + 2\sqrt{10})$.
18. $s^2 = 4s$; $s = 4$.
19. Area of a triangle $= \frac{1}{2}bh$.

    Area of a trapezoid $= \frac{1}{2}h(b + b_2)$

    $$\frac{1}{2}bh + \frac{1}{2}b_2h > \frac{1}{2}bh.$$
20. $2x + 2y = 24$; $x + y = 12$; $y = 12 - x$; $A = bh = x(12 - x) = 20$; $x^2 - 12x + 20 = (x - 10)(x - 2) = 0$; the dimensions are 10 and 2.

Now, let's look at circles.

Answer to "Bob Asks": It is the same: 4% less. To show that this is true, start with \$100; an increase of 20% gives \$120, and 20% of this is \$24. So the net is \$96. Now start with \$100 and decrease it by 20%, which is \$80; 20% of \$80 is \$16, so we get \$80 + \$16 = \$96.

CHAPTER 12: *Circles, Circles, Circles*

"An octagon has a perimeter of 5 feet. If 3 inches are added to each side, what is the perimeter of the new octagon?"

Circles allow for many short questions that can be combined with the other geometric shapes. Let's get started.

PARTS OF A CIRCLE

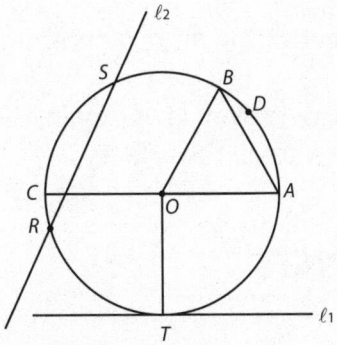

We all know what a circle looks like, but maybe we're not familiar with its "parts."

O is the **center** of the circle. *A* circle is often named by its center, so this is circle *O*.

OA, *OT*, *OC*, and *OB* are **radii** (singular: **radius**); a radius is a line segment from the center to the **circumference**, or edge, of the circle.

Note *All radii (r) of a circle are equal. This is a postulate or axiom, a law taken to be true without proof.*

It is a good idea to tell you there are no proofs on the ACCUPLACER, as there probably were when you took geometry.

AC is the **diameter**, *d*, the distance from one side of the circle through the center to the other side; $d = 2r$ and $r = \frac{d}{2}$.

ℓ_1 is a **tangent**, a line that touches a circle in one and only one point, *T* here.

T is a **point of tangency**, the point where a tangent touches the circle. The radius to the point of tangency (*OT*) is always perpendicular to the tangent, so $OT \perp \ell_1$.

ℓ_2 is a **secant**, a line that passes through a circle in two places.

RS is a **chord**, a line segment that has each end on the circumference of the circle. The diameter is the longest chord in a circle.

OADBO is a **sector** (a pie-shaped part of a circle). There are a number of sectors in this figure; others include *BOCSB* and *OATRCSBO*. We will see these again soon.

An **arc** is any distance along the circumference of a circle.

> Arc *ADB* is a **minor arc** because it is less than half a circle. Arc *BDATRC* is a **major arc** because it is more than half a circle.
>
> Arc *ATRC* is a **semicircle** because it is exactly half a circle. Arc *ADBSC* is also a semicircle.

Whew! Enough! However, we do need some more facts about circles.

The following are mostly theorems, or proven laws. Again, there are no proofs on the ACCUPLACER, but you need to be aware of these facts.

AREA AND CIRCUMFERENCE

Area of a circle:

$$A = \pi r^2$$

Circumference (perimeter of a circle):

$$C = 2\pi r \text{ or } \pi d$$

SECTORS

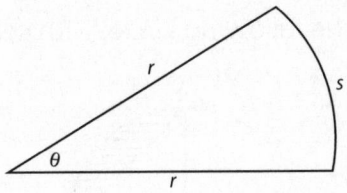

Area of a sector:

$$A = \frac{\theta}{360^\circ}\pi r^2,$$

where θ (theta) is the angle of the sector in degrees.

Arc length of a sector:

$$s = \frac{\theta}{360^\circ} \times 2\pi r$$

Perimeter of a sector:

$$p = s + 2r,$$

where s is the arc length.

Example 1: Find the area and perimeter of a 60° sector of a circle of diameter 12.

Solution:

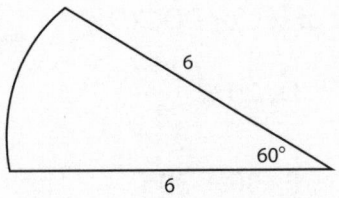

If the diameter is 12, the radius is 6. The sector is pictured above. Its area is $A = \frac{60^\circ}{360^\circ} \times \pi 6^2 = 6\pi$ square units. Although you should know that pi (π) is about 3.14, I've never seen a problem for which you had to multiply 6 times 3.14. The answer is left in terms of π. The perimeter of the sector is $p = 2(6) + \frac{60^\circ}{360^\circ} \times 2\pi(6) = 12 + 2\pi$ units.

Q Let's do some multiple-choice exercises.

For Exercises 1 through 5, refer to the following circle, with center *O*.

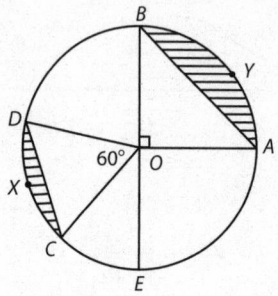

Exercise 1: If the area of ΔAOB is 25, the area of circle *O* is

A. 12.5π
B. 25π
C. 50π
D. 100π
E. 200π

Exercise 2: If $OA = 8$, the area of the shaded region *BAYB* is

A. $16(\pi - 2)$
B. $16(\pi - 1)$
C. $8(2\pi - 1)$
D. $64(\pi - 1)$
E. $32(\pi - 2)$

Exercise 3: If $CD = 10$, the perimeter of sector *DOCXD* is

A. $30 + \frac{10\pi}{3}$
B. $30 + \frac{20\pi}{3}$
C. $20 + \frac{10\pi}{3}$
D. $20 + \frac{20\pi}{3}$
E. $30 + 30\pi$

Exercise 4: If $OC = 2$, the area of the shaded portion *DCXD* is

A. $\pi - \sqrt{3}$
B. $2\pi - \sqrt{3}$
C. $4\pi - \sqrt{3}$
D. $\frac{2\pi - 3\sqrt{3}}{3}$
E. $\frac{8\pi - 3\sqrt{3}}{3}$

Exercise 5: If the area of ΔCOD is $25\sqrt{3}$, the perimeter of semicircle *EOBDXCE* is

A. 10π

B. $10(\pi + 1)$

C. $10(\pi + 2)$

D. $10(2\pi + 1)$

E. $20(\pi + 1)$

I guess you get the idea already.

Let's look at the answers.

Answer 1: **C:** $A = \frac{1}{2}(r)(r) = \frac{1}{2}r^2 = 25$, so $r^2 = 50$. The area of the circle is thus $\pi r^2 = 50\pi$. Notice that once we have a value for r^2, we don't have to find r to do this problem.

Answer 2: **A:** The area of region *BAYB* is the area of one-fourth of a circle minus the area of ΔAOB. So the area is $A = \frac{1}{4}\pi 8^2 - \frac{1}{2}(8)(8) = 16\pi - 32 = 16(\pi - 2)$.

Answer 3: **C:** The perimeter of sector *DOCXD* $= 2r + s$, where s is the length of arc *CXD*. ΔCOD is equilateral, so $CD = CO = DO = r = 10$. $s = \frac{60°}{360°}2\pi(10) = \frac{10\pi}{3}$. So the perimeter of sector *DOCXD* is $p = 20 + \frac{10\pi}{3}$.

Answer 4: **D:** The area of region *DCXD* is the area of sector *ODXCO* minus the area of ΔDOC, when $OC = 2$. So the area is

$$A = \frac{60°}{360°}\pi 2^2 - \frac{2^2\sqrt{3}}{4} = \frac{2\pi}{3} - \sqrt{3} = \frac{2\pi - 3\sqrt{3}}{3}.$$

Answer 5: **C:** We must first find the radius. The area of equilateral $\Delta COD = \frac{s^2\sqrt{3}}{4} = 25\sqrt{3}$. So $s^2 = 100$, and $s = r = 10$. The perimeter of the semicircle is $\frac{1}{2}(2\pi r) + 2r = \pi r + 2r = 10\pi + 20 = 10(\pi + 2)$.

There are a few more things we need to know. When we talked about two intersecting line segments earlier, we saw that, for the figure below, *CE* might equal *ED*; however, if the description of the figure doesn't say so, you cannot assume it. Also, *ED* might be perpendicular to *AB*, but if it doesn't say so, you cannot assume it, either. In fact, we can say *CD* **bisects** *AB* at *E* only if we know that $AE = EB$ or *E* is the **midpoint** of *AB*.

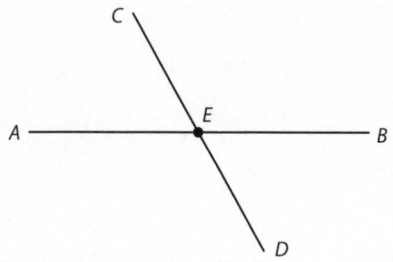

Now, however, we consider two intersecting line segments in a circle, such as chord *AB* and radius *CO* in circle *O*.

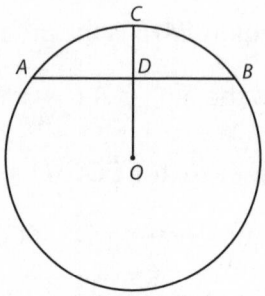

If one of the following facts is true, all are true:

1. $OD \perp AB$
2. *CO* bisects *AB at point D.*
3. *CO* bisects $\overset{\frown}{ACB}$ (read "arc *ACB*")

Note *When the ACCUPLACER says "distance to a chord," distance always means* ***perpendicular*** *distance. Sometimes the ACCUPLACER uses the right-angle sign to show lines are perpendicular (see the figure for Exercise 6). When it doesn't, you have to know it is perpendicular from the information given.*

Let's do some more multiple-choice exercises.

For Exercises 6 and 7, use this figure, which is a triangle-semicircle shape.

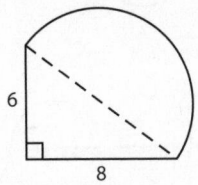

Exercise 6: The perimeter of this figure is

A. $14 + 5\pi$
B. $14 + 10\pi$
C. $24 + 5\pi$
D. $24 + 10\pi$
E. $12 + 10\pi$

Exercise 7: The area of this figure is

A. $24 + \frac{25\pi}{2}$
B. $24 + 25\pi$
C. $24 + 50\pi$
D. $48 + 25\pi$
E. $48 + 50\pi$

Exercise 8: A circle is **inscribed** in (inside and touching) figure *MNPQ*, which has all right angles. Diameter $AB = 10$. The area of the shaded portion is

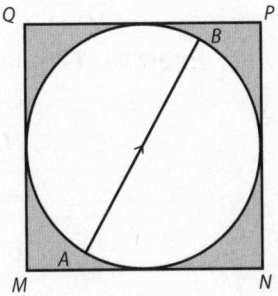

A. $100 - 12.5\pi$
B. $100 - 25\pi$
C. $40 - 10\pi$
D. $40 - 5\pi$
E. $100 - 100\pi$

Exercise 9: If $AB = 10$, the area of the shaded portion in the figure is

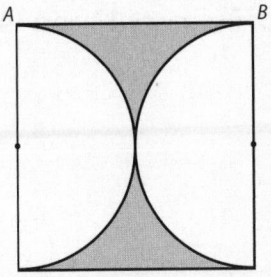

A. $100 - 12.5\pi$
B. $100 - 25\pi$
C. $40 - 10\pi$
D. $40 - 5\pi$
E. $100 - 100\pi$

For Exercises 10 and 11, use this figure. The perimeter of the 16 semicircles is 32π.

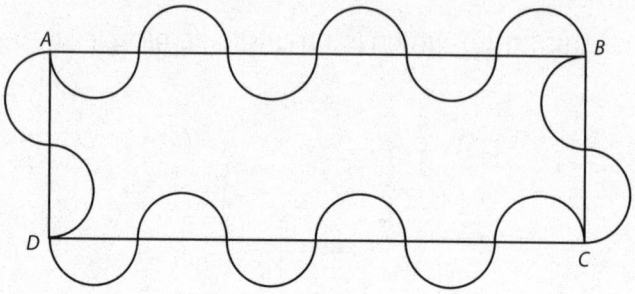

Exercise 10: The area of rectangle *ABCD* is

A. 64

B. 128

C. 192

D. 16π

E. Cannot be determined

Exercise 11: The area inside the region formed by the semicircular curves from *A* to *B* to *C* to *D* and back to *A* is

A. 64

B. 128

C. 192

D. 16π

E. Are you for real??!!

Exercise 12: In the figure, $EF = CD = 12$, *B* is the midpoint of *OD*, and *A* is the midpoint of *CO*. The area of the shaded portion is

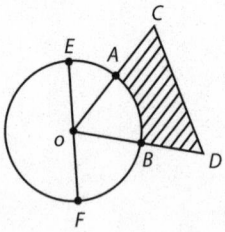

A. $36(4\sqrt{3} - \pi)$

B. $6(6\sqrt{3} - \pi)$

C. $144(\pi - 3)$

D. $72(3\sqrt{3} - \pi)$

E. $36(6\sqrt{3} - \pi)$

For Exercises 13 and 14, use this figure.

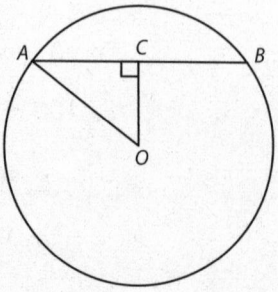

Exercise 13: If the diameter $= 20$, $OC \perp AB$, and $\angle A = 30°$, $AB =$

A. 10
B. 5
C. $5\sqrt{3}$
D. $10\sqrt{3}$
E. $5\sqrt{2}$

Exercise 14: If OC bisects AB, $AB = 16$, and $OC = 6$, the area of circle O is

A. 10π
B. 20π
C. 40π
D. 100π
E. 400π

Let's look at the answers.

Answer 6: **A:** The figure includes a 6-8-10 Pythagorean triple, but $d = 10$ is not part of the perimeter. However, we need the radius $r = \frac{1}{2}d = 5$ for the circle. So $p = 6 + 8 + \frac{1}{2}2\pi(5) = 14 + 5\pi$.

Answer 7: **A:** $A = \frac{1}{2}bh + \frac{1}{2}\pi r^2 = \frac{1}{2}6 \times 8 + \frac{1}{2}\pi 5^2 = 24 + \frac{25\pi}{2}$.

Answer 8: **B:** The area is the area of the square minus the area of the circle. $A = s^2 - \pi r^2 = 10^2 - \pi 5^2 = 100 - 25\pi$.

Answer 9: **B:** Exercises 8 and 9 are exactly the same problems. In Exercise 8, we could also say the square **circumscribes** the circle.

Answer 10: **C:** Each semicircle has arc length $\frac{180°}{360°}\pi d$, and there are 16 semicircles, so $16\left(\frac{1}{2}\pi d\right) = 32\pi$, or $8\pi d = 32\pi$, so $d = 4$. The rectangle's dimensions are thus 8 and 24. The area is $A = b \times h = 8(24) = 192$.

Answer 11: **C:** Believe it or not, Exercise 11 is exactly the same as Exercise 10! We can think of the areas of the "outer" semicircles as canceling out the areas of the "inner" semicircles, and we are left with only the area of rectangle $ABCD$.

Answer 12: **B:** The information is enough to tell us the triangle is equilateral and $\angle AOB = 60°$. The shaded area is the area of ΔCOD minus the area of sector ABO. Thus, $A = \frac{s^2\sqrt{3}}{4} - \frac{1}{6}\pi r^2 = \frac{12^2\sqrt{3}}{4} - \frac{1}{6}\pi 6^2 = 36\sqrt{3} - 6\pi = 6(6\sqrt{3} - \pi)$.

Answer 13: **D:** AO, the radius, is 10; CO, the side opposite the 30° angle, is 5; and AC, the side opposite the 60° angle, is $5\sqrt{3}$. $AB = 2(AC) = 2(5\sqrt{3}) = 10\sqrt{3}$.

Answer 14: **D:** To find the area of the circle, we need to find the radius OA. We know AC is 8 and OC is 6. We have a 6-8-10 right triangle, so $AO = r = 10$. The area is $\pi(10)^2 = 100\pi$.

Chapter 12 Test

All answers may be left in terms of π.

1. A line segment from one side of the circle to the other is called what?
2. A line hitting a circle in two places is called what?
3. What is the part of the circle that is larger than a semicircle?
4. The point at which the tangent line touches the circle is called what?
5. A line segment from the center of a circle to the circle is perpendicular to a chord. What other two facts are true about this line segment?
6. Find the area of a circle if the diameter is 14.
7. Find the circumference of a circle if the radius is 7.6.
8. What is the diameter of the circle if the area and circumference are the same number?

Questions 9–11 refer to a circle with a 20° sector and a diameter of 20.

9. Find the area of the sector.
10. Find the length of the arc.
11. Find the perimeter of the sector.

Questions 12 and 13 refer to a Norman window (a rectangle with a semicircle on top). The semicircle is on top of the side of the rectangle that is $4x$; the other side of the rectangle is $8x$.

12. Find the perimeter of the Norman window in terms of x and π.
13. Find the area of the Norman window in terms of x and π.

Questions 14 and 15 refer to two concentric circles (having the same center).

14. If the inner diameter is 14 and the outer diameter is 16, find the area between the circles.
15. If the inner radius is 10 and the width between the circles is 2, find the area between the circles.
16. A circle is inside a square, touching the square in 4 points. The side of the square is 6. Find the area inside the square but outside the circle.
17. A square is inside a circle, touching the circle in 4 points. The side of the square is 6. Find the area inside the circle and outside the square.
18. A circle has area 20π. If a sector has area 4π, how many degrees are in the central angle of the sector?
19. A regular hexagon is inscribed in a circle of diameter 20. Find its area.
20. A line segment of length 6 goes from the center of a circle, perpendicular to a chord of length 16. Find the area of the circle.

Answers to Chapter 12 Test

1. Chord.
2. Secant.
3. Major arc.
4. Point of tangency.
5. It bisects the chord and it bisects the arc.
6. 49π.
7. 15.2π.
8. $\pi r^2 = 2\pi r$; $r = 2$; so $d = 4$.
9. $\frac{20}{360} \times \pi \times 10^2 = \frac{50}{9}\pi$.
10. $\frac{20}{360} \times 2\pi \times 10 = \frac{10\pi}{9}$.
11. $20 + \frac{10\pi}{9}$.
12. $8x + 8x + 4x + \left(\frac{1}{2}\right)2\pi(2x) = 20x + 2\pi x$.
13. $(4x)(8x) + \left(\frac{1}{2}\right)\pi(2x)^2 = 32x^2 + 2\pi x^2$.

14. $\pi\left(r^2_{\text{outside}} - r^2_{\text{inside}}\right) = \pi(8^2 - 7^2) = 15\pi.$

15. $\pi(12^2 - 10^2) = 44\pi.$

16. $r = 3; A = 36 - 9\pi.$

17. $r = \dfrac{6}{\sqrt{2}} = 3\sqrt{2}; A = 18\pi - 36.$

18. $\dfrac{4\pi}{20\pi} = \dfrac{1}{5}$ the circle $= 72°$.

19. A regular hexagon is six equilateral triangles where $s = r = 10$;

    $A = 6 \times 10^2 \dfrac{\sqrt{3}}{4} = 150\sqrt{3}.$

20. The segment bisects the chord (8). The picture is a 6–8–10 right triangle, where the radius is 10. The area of the circle is 100π.

Okay. Now let's go from two dimensions to three dimensions.

Answer to "Bob Asks": 7 feet. It is not necessary to know how long each side is. Adding 3 inches to each of 8 sides means 24 inches, or 2 feet, are added to the original perimeter.

CHAPTER 13: *All About Three-Dimensional Figures*

"A square has a perimeter of 1 foot. If we have a square with each side half as large, what is the perimeter of the smaller square?"

This chapter is relatively short. There are only a few figures we need to know. Because these are three-dimensional figures, we discuss their volumes and surface areas (areas of all of the surfaces). The diagonal is the distance from one corner internally to an opposite corner.

BOX

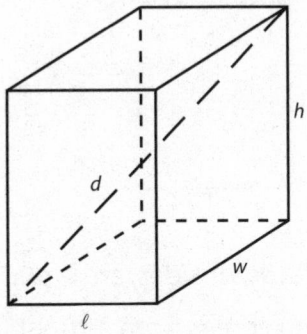

This figure is also known as a **rectangular solid**, and if that isn't a mouthful enough, its correct name is a **rectangular parallelepiped**. But essentially, it's a **box**.

- Volume $= V = \ell wh$
- Surface area $= SA = 2\ell w + 2\ell h + 2wh$
- Diagonal $= d = \sqrt{\ell^2 + w^2 + h^2}$, known as the 3-D Pythagorean Theorem

Example 1: For the given figure, find V, SA, and d.

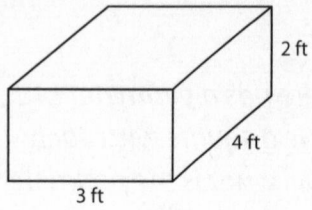

Solution: $V = \ell wh = (3)(4)(2) = 24$ cubic feet; $SA = 2\ell w + 2\ell h + 2wh = 2(3)(4) + 2(3)(2) + 2(4)(2) = 52$ square feet; $d = \sqrt{\ell^2 + w^2 + h^2} = \sqrt{3^2 + 4^2 + 2^2} = \sqrt{29} \approx 5.4$ feet

CUBE

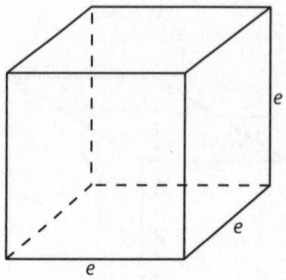

A **cube** is a box for which all of the faces, or sides, are equal squares.

- $V = e^3$ (read as "e cubed"). Cubing comes from a cube!
- $SA = 6e^2$
- $d = e\sqrt{3}$
- A cube has 6 faces, 8 vertices, and 12 edges.

Example 2: For a cube with an edge of 10 meters, find V, SA, and d.

Solution: $V = 10^3 = 1{,}000$ cubic meters; $SA = 6e^2 = 6(10)^2 = 600$ square meters; $d = e\sqrt{3} = 10\sqrt{3} \approx 17.32$ meters

CYLINDER

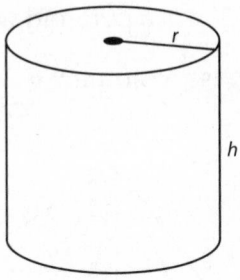

A cylinder is shaped like a can. The curved surface is considered as a side, and the top and bottom are equal circles.

- $V = \pi r^2 h$
- SA = top + bottom + curved surface = $2\pi r^2 + 2\pi rh$

Once a neighbor of mine wanted to find the area of the curved part of a cylinder. He wasn't interested in why, just the answer. Of course, being a teacher, I had to explain it to him. I told him that if he cut a label off a soup can and unwrapped it, the figure is a rectangle; neglecting the rim, the height is the height of the can and the width is the circumference of the circle. Multiply this height and width, and the answer is $2\pi r \times h$. He waited patiently and then soon moved. (Just kidding!)

In general, the volume of any figure for which the top is the same as the bottom is $V = Bh$, where B is the area of the base. If the figure comes to a point, the volume is $\left(\frac{1}{3}\right)Bh$. The surface area is found by adding up all the surfaces.

Example 3: Find V and SA for a cylinder of height 10 yards and diameter of 8 yards.

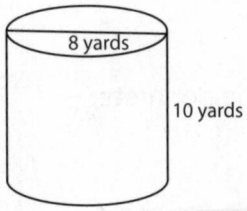

Solution: We see that, because $d = 8$, $r = 4$. Then $V = \pi r^2 h = \pi(4^2 \times 10) = 160\pi$ cubic yards; $SA = 2\pi r^2 + 2\pi rh = 2\pi 4^2 + 2\pi(4)(10) = 112\pi$ square yards.

Q Let's do some multiple-choice exercises.

Use this figure for Exercises 1 through 3. It is a pyramid with a square base. $WX = 8$, $BV = 3$, and pyramid B is in the middle of the base. Volume $V = \frac{1}{3}$ area of base $\times$ height.

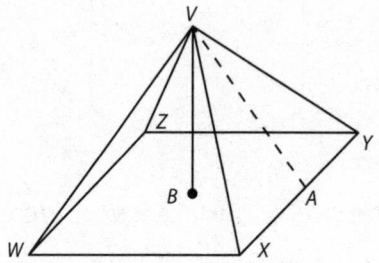

Exercise 1: The volume of the pyramid is

A. 192
B. 96
C. 64
D. 32
E. 16

Exercise 2: The surface area of the pyramid is

A. 72
B. 112
C. 144
D. 224
E. 448

Exercise 3: $VY =$

A. 6
B. $\sqrt{41}$
C. 7
D. 9
E. $\sqrt{89}$

Exercise 4: In the given rectangular solid, the perimeter of $\Delta ABC =$

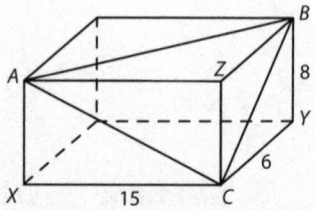

A. $\sqrt{325} = 5\sqrt{13}$
B. 30
C. $27 + \sqrt{261}$
D. 37
E. 41

Exercise 5: The volume of the cylinder shown is:

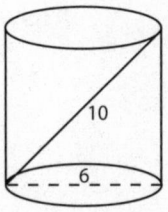

A. 640π

B. 320π

C. 288π

D. 144π

E. 72π

Exercise 6: *ABKL* is the face of a cube with $AB = 10$, and box *BCFG* has a square front with $BC = 6$. The surface area that can be viewed in this configuration is

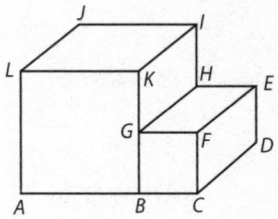

A. 300

B. 356

C. 396

D. 400

E. 1360

Exercise 7: A cylinder has volume *V*. If we triple its radius, what do we multiply the height by in order for the volume to stay the same?

A. $\frac{1}{9}$

B. $\frac{1}{3}$

C. 1

D. 3

E. 9

Let's look at the answers.

Answer 1: **C:** $V = \left(\frac{1}{3}\right)Bh = \frac{1}{3}s^2h = \frac{1}{3}(8^2)(3) = 64.$

Answer 2: **C:** $SA = s^2 + 4\left(\frac{1}{2}bh\right)$. $AB = \left(\frac{1}{2}\right)WX = 4$; ΔABV is a 3-4-5 right triangle with $AB = 4$ and $BV = 3$, so $AV = 5$. AV is the height of each triangular side, h, and $b = XY = 8$. So $SA = 8^2 + 2(8)(5) = 144$.

Answer 3: **B:** ΔAVY is a right triangle with right angle at A. $AY = 4$ and $AV = 5$, so $VY = \sqrt{4^2 + 5^2} = \sqrt{41}$.

Answer 4: **C:** In the given figure, we have to use the 2-D Pythagorean Theorem three times to find the sides of ΔABC. ΔBCY is a 6-8-10 triple, so $BC = 10$, and ΔACX is a 8-15-17 triple, so $AC = 17$. For ΔABZ, we actually have to calculate the missing side $AB = \sqrt{6^2 + 15^2} = \sqrt{261}$. So the perimeter is $10 + 17 + \sqrt{261} = 27 + \sqrt{261}$.

Answer 5: **E:** The diameter of the base is 6, and again we have a Pythagorean 6-8-10 triple; so $h = 8$. The volume is $\pi(3^2)(8) = 72\pi$.

Answer 6: **C:** The areas are: $ABKL = 100$; $IJLK = 100$; $BCFG = 36$; $CDEF = 60$; $EFGH = 60$; $GHIK = 40$. The total is 396.

Answer 7: **A:** The volume $V = \pi r^2 h$. For simplicity, let $r = 1$ and $h = 1$. So $V = \pi$. If we triple the radius, $V = \pi(3)^2 h$. For the original volume to still be π, $9h = 1$, or $h = \frac{1}{9}$.

SPHERE

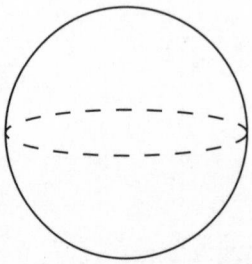

Another figure we need to know the formulas for is the **sphere,** on which all points are the same distance from the center. That distance is called the **radius** r of the sphere.

- $V = \frac{4}{3}\pi r^3$
- $SA = 4\pi r^2$

Example 4: If a sphere has radius of 10 inches, find its volume and surface area.

Solution: $V = \frac{4}{3}\pi r^3 = \frac{4000\pi}{3}$ cubic inches. $SA = 4\pi r^2 = 400\pi$ square inches.

CONE

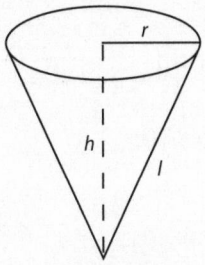

The last 3D shape you need to know is the **cone;** technically, this is a right circular cone. The word "right" here means the height is to the middle of the base, which is a circle. The **slant height,** which is the distance from any point on the circle to the point at the top of the cone, is designated by ℓ.

- $V = \frac{1}{3}\pi r^2 h$
- $SA = \pi r^2 + \pi r\ell$
- $\ell = \sqrt{r^2 + h^2}$

Example 5: Find the volume and surface area of a cone with $r = 8$ and $h = 7$.

Solution: We need to find the slant height first. The slant height is

$\ell = \sqrt{8^2 + 7^2} = \sqrt{113}$. $V = \frac{1}{3}\pi(8^2)(7) = \frac{448\pi}{3}$ and

$SA = \pi 8^2 + \pi(8)\sqrt{113} = 8\pi(8 + \sqrt{113})$.

Note *Every formula involving π has been derived with calculus. One thing that thrills many students is to be actually shown that the area of a circle is πr^2! It really is true!*

Chapter 13 Test

Answers may be left in terms of π.

Questions 1 and 2 refer to a sphere of diameter 18.

1. Find the volume of the sphere.
2. Find the surface area of the sphere.

Questions 3–5 refer to a box with width 5, length 6, and height 7.

3. Find the volume of the box.
4. Find the surface area of the box.
5. Find the diagonal of the box.

Questions 6 and 7 refer to a cylinder of diameter 10 and height 20.

6. Find the volume of the cylinder.
7. Find the surface area of the cylinder.

Questions 8–10 refer to a cube with edge 10.

8. Find the volume of the cube.
9. Find the surface area of the cube.
10. Find the diagonal of the cube.

Questions 11–13 refer to a cone of diameter 6 and height 4.

11. Find the slant height of the cone.
12. Find the volume of the cone.
13. Find the surface area of the cone.
14. If the volume of a cube is numerically equal to its surface area, what is the length of a side?
15. For what radius is the volume of a sphere equal numerically to its surface area?

Questions 16 and 17 refer to a child's toy block that is a cylinder with a hemisphere (half a sphere) on top of it.

16. Find its surface area if the radius = 8 and the height = 10.
17. Find its volume if the radius = 8 and the height = 10.
18. A pyramid has a rectangular base that is 5 feet by 3 feet with a height of 7 feet. Find its volume.
19. For a cube, verify that the number of sides plus the number of vertices equals the number of edges plus 2.
20. What is the minimum number of colors you need to color a cube so that the same colors don't touch?

Answers to Chapter 13 Test

1. $r = 9; V = \frac{4}{3}\pi \times 9^3 = 972\pi.$
2. $S = 4\pi \times 9^2 = 324\pi.$
3. $V = (5)(6)(7) = 210.$
4. $S = 2[(6)(7) + (5)(7) + (5)(6)] = 214.$
5. Diagonal $= \sqrt{5^2 + 6^2 + 7^2} = \sqrt{110}.$

6. $V = \pi(5)^2 20 = 500\pi$.
7. $S = 2\pi(5)^2 + 2\pi(5)(20) = 250\pi$.
8. 1000.
9. 600.
10. $10\sqrt{3}$.
11. The slant height is 5 because $r = 3$, and it is the hypotenuse of a 3-4-5 right triangle.
12. $V = \frac{1}{3}\pi(3)^2 4 = 12\pi$.
13. $SA = \pi(3)^2 + \pi(3)(5) = 24\pi$.
14. 6.
15. 3.
16. $SA = \pi r^2 + 2\pi(r)(h) + \left(\frac{1}{2}\right)4\pi r^2 = (64 + 160 + 128)\pi = 352\pi$.
17. $V = \pi r^2 h + \frac{1}{2} \times \frac{4}{3}\pi r^3 = \pi\left(640 + \frac{1024}{3}\right) = \left(981\frac{1}{3}\right)\pi$.
18. $\frac{1}{3}(5)(3)(7) = 35$.
19. $6 + 8 = 12 + 2$.
20. Three! The top and bottom are one color; the front and the back are another color; and the two sides are a third color. This one is for fun. Yes, in math some questions are fun questions, at least for me and hopefully for you.

Enough for geometry! Let's move on to other topics you might see on the ACCUPLACER.

Answer to "Bob Asks": $\frac{1}{2}$; half the sides means half the perimeter. Oh, let's show it. The perimeter of the first square is 1 foot, which means the sides are 3 inches each. Half of that is $1\frac{1}{2}$ inches; times 4 is 6 inches, or $\frac{1}{2}$ foot.

CHAPTER 14: *More Algebraic Topics*

"A monkey is trying to climb out of a 50-foot hole. In the morning, it climbs 5 feet. At night, it gets so tired that it falls back 4 feet. The monkey does this every day. On what day does the monkey get out of the hole?"

FUNCTIONS

A most important part of algebra is the study of functions. Let's give it its due.

Function: To each element in set D, we assign one and only one element. The set D is called the **domain**. On the ACCUPLACER, you should think of the x values. The assignment is called the **map**; the set of numbers that are assigned is called the **range**. On the ACCUPLACER, you should think of the y values.

Note *If you are unfamiliar with sets and elements, see Chapter 16.*

Example 1: Let $f(x) = x^2 + 3x + 22$; find the following:

Problem	Solution
a. $f(4)$	$4^2 + 3(4) + 22 = 50$
b. $f(-3)$	$(-3)^2 + 3(-3) + 22 = 22$
c. $f(0)$	$0^2 + 3(0) + 22 = 22$
d. $f(x + h)$	$(x + h)^2 + 3(x + h) + 22 =$ $x^2 + 2xh + h^2 + 3x + 3h + 22$
e. $f(\text{pigs})$	$(\text{pigs})^2 + 3\text{pigs} + 22$

Example 2: Let $g(x) = x^2 + 7x$; let $D = \{-3, 4, 7\}$. Find the range.

Solution: Range $= \{g(-3), g(4), g(7)\} = \{-12, 44, 98\}$

Note *Functions (the maps) are usually indicated by f(x), g(x), F(x), and G(x). However, any letter can be used.*

Note *A graph is a function if any time you draw a vertical line, the line crosses the graph only once. All lines (except a vertical line) are functions; a circle is not a function.*

PIECEWISE FUNCTIONS

Sometimes a function has more than one piece. Let's give an example; we'll do it the longer way for understanding, then the shorter way, and finally how the ACCUPLACER might ask the question.

Example 3: Graph $f(x) = \begin{cases} x + 6 & \text{if } x < 0 \\ x^2 & \text{if } 0 \le x \le 3 \\ 5 - x & \text{if } x > 3 \end{cases}$

Solution: We make a table of values: $x = -3, -2, -1, 0^-, 0, 1, 2, 3, 3^+, 4, 5, \ldots$

When $x < 0$, we use the first part of the definition, $f(-3) = (-3) + 6 = 3$, and we have the point $(-3, 3)$.

Similarly, $f(-2) = (-2) + 6 = 4$, and we have the point $(-2, 4)$; $f(-1) = (-1) + 6 = 5$, which yields the point $(-1, 5)$; $f(0^-) = 0^- + 6 = 6^-$ (or a little less than 0, like $-.0001$, plus 6 equals a little less than 6). The point is $(0^-, 6^-)$—on a graph this is an open dot because the graph doesn't quite hit $(0, 6)$.

For the second part of the definition, $f(0) = 0^2 = 0$ yields the point $(0, 0)$; $f(1) = 1^2 = 1$ gives $(1, 1)$; $f(2) = 2^2 = 4$ gives $(2, 4)$; and $f(3) = 9$ gives $(3, 9)$.

For the third part of the definition, $f(3^+) = 5 - 3^+ = 2^-$ (5 minus a little more than 3 equals a little less than 2, or 2^-), which yields the point $(3^+, 2^-)$ with an open dot on $(3, 2)$; $f(4) = 5 - 4 = 1$ gives $(4, 1)$; and $f(5) = 5 - 5 = 0$ gives $(5, 0)$.

The graph would look like this:

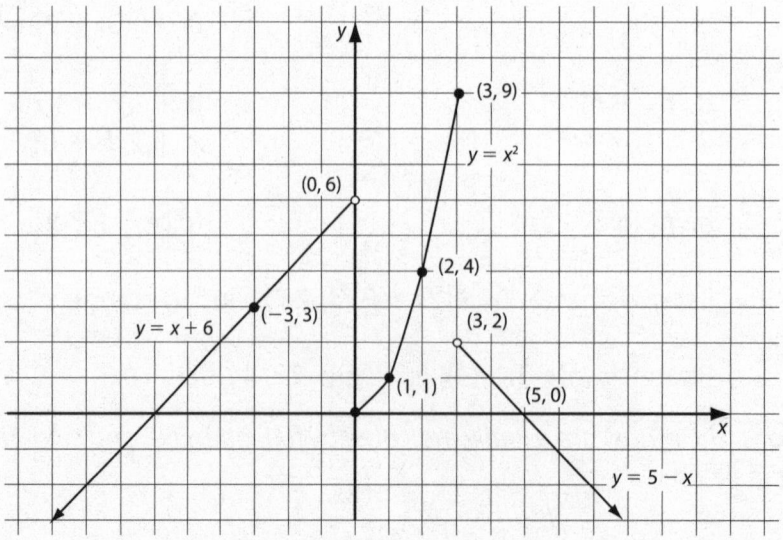

To be able to do this same problem the short way, you must know or be able to get the graph really quickly.

We graph $y = x + 6$, but we are interested only in the part of the line where $x < 0$.

We graph $y = x^2$, but we are interested only in the part of the graph between 0 and 3, inclusive.

We graph $y = 5 - x$, but we are interested only in the part of the graph where $x > 3$.

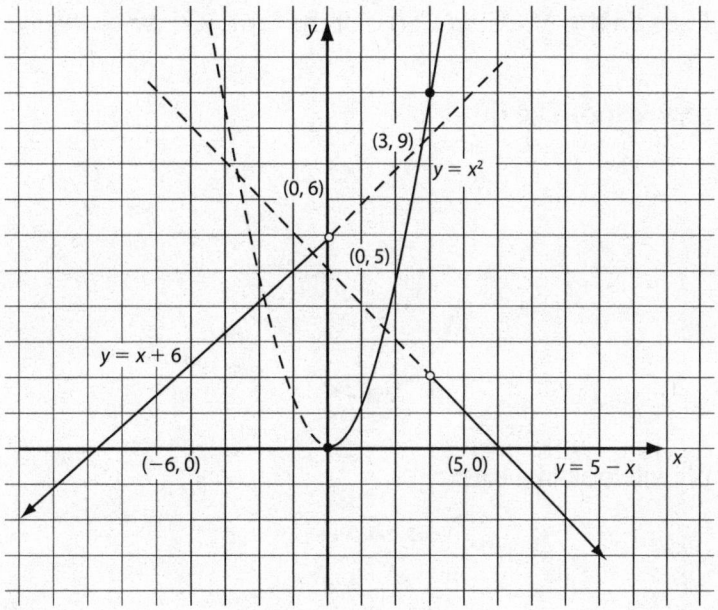

The ACCUPLACER could provide a value for $f(x)$ and you would have to know what part of the definition of a piecewise function holds and find x.

The ACCUPLACER will have a question with five answer choices, one of which tells what x is.

Let's look at a multiple-choice exercise.

Exercise 1: Referring to the figure above, if $f(x) = 4$, x could be

A. 2 only

B. −2 only

C. 2, −2 only

D. 1 and 2 only

E. 1, 2, and 4 only

Let's look at the answer.

Answer 1: **C:** If $f(x) = x + 6 = 4$; $x = -2$; because $-2 < 0$, $x = -2$ is okay. If $f(x) = x^2 = 4$; $x = 2$ and $x = -2$; only 2 is between 0 and 3; so $x = 2$ is okay. If $f(x) = 5 - x = 4$, $x = 1$; because 1 is not > 3, $x = 1$ is not okay.

PARABOLAS

The parabola you will see on the ACCUPLACER is of the form $y = ax^2 + bx + c$, where a cannot be 0.

- If $a > 0$, the graph will look like this:

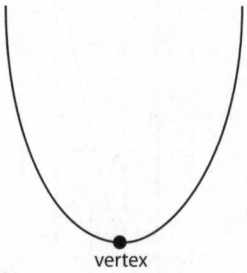

- If $a < 0$, the graph will look like this:

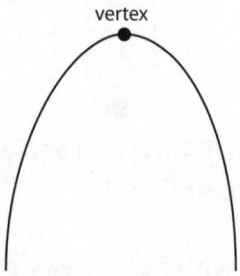

The vertex is the lowest or highest point of the parabola. Its x value is found by setting $x = -\frac{b}{2a}$.

Example 4: Sketch $y = x^2 - 4x - 5$.

Solution: We sketch by finding the vertex, x-intercept(s) and y-intercept.

Vertex: $x = -\frac{b}{2a} = -\frac{-4}{2} = 2$; $y = (2)^2 - 4(2) - 5 = -9$: $(2, -9)$

x-intercept(s): $y = 0$; $x^2 - 4x - 5 = (x - 5)(x + 1) = 0$: $(5, 0)$ and $(-1, 0)$.

y-intercept: $x = 0$; $y = -5$: $(0, -5)$.

The sketch looks like this:

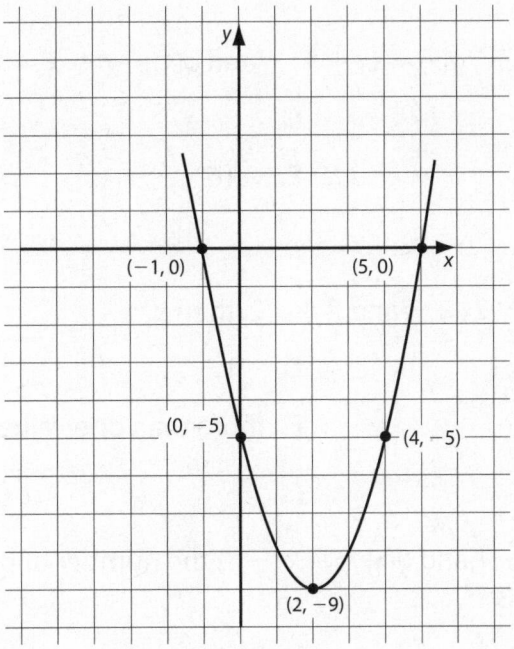

Example 5: Sketch $y = 18 - 2x^2$.

Solution: We sketch by finding the vertex, x-intercept(s), and y-intercept.

Vertex: $x = -\dfrac{b}{2a} = -\dfrac{0}{2(2)} = 0$; $y = 18$: (0, 18). It is also the y-intercept.

x-intercepts: $y = 0$; $18 - 2x^2 = -2(x^2 - 9) = -2\,(x + 3)(x - 3) = 0$:

(−3, 0) and (3, 0).

The sketch looks like this:

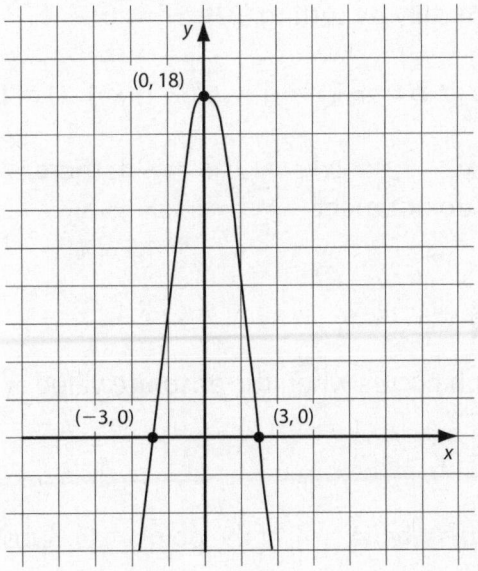

Note *With no x term, the parabola is symmetric with respect to the y-axis.*

Q Let's try a few multiple-choice exercises:

Exercise 2: $f(x) = (x + 5)^2$, $g(x) = (x + 11)^2$, and $g(x) = f(x)$; $x =$

A. 16
B. 8
C. 0
D. -8
E. -16

Exercise 3: If $f(x) = x^2 + 2x + 3$ and $f(x) = 2$, then $x =$

A. 11
B. 1
C. 0
D. -1
E. more than one value

Exercise 4: If $f(x) = x^2 + 1$ and $g(x) = -x^2 - 1$; the number of points where $f(x)$ meets $g(x)$ is

A. 0
B. 1
C. 2
D. 3
E. more than 3

A Let's look at the answers.

Answer 2: **D:** The answer is halfway between the two; halfway between -5 and -11 is -8. You could see this if you sketched the curve. Or you could solve it algebraically by setting $(x + 5)^2 = (x + 11)^2$.

Answer 3: **D:** $x^2 + 2x + 3 = 2$; $x^2 + 2x + 1 = (x + 1)(x + 1) = 0$; so $x = -1$.

Answer 4: **A:** $x^2 + 1 = -x^2 - 1$; so $2x^2 = -2$; $x^2 = -1$; there is no real solution, so the graphs do not meet.

ABSOLUTE VALUE GRAPH

The vertex of an absolute value graph occurs when the absolute value $= 0$.

Example 6: Let's graph $y = |x - 4|$.

Solution: The vertex occurs when $x = 4$, at the point (4, 0). Substituting $x = 2, 3, 5,$ and 6, we get the following graph:

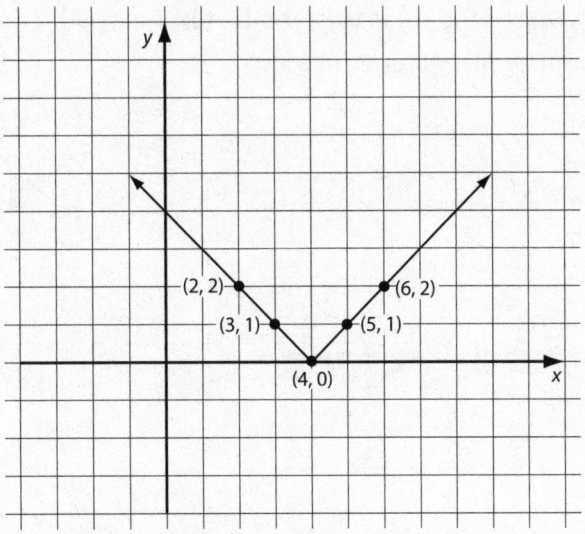

Example 7: Let's graph $y = -|x + 3| + 1$.

Solution: The vertex occurs at $x = -3$; $y = 1$; $(-3, 1)$. Substituting $x = -5, -4, -2, -1$, we get the following graph:

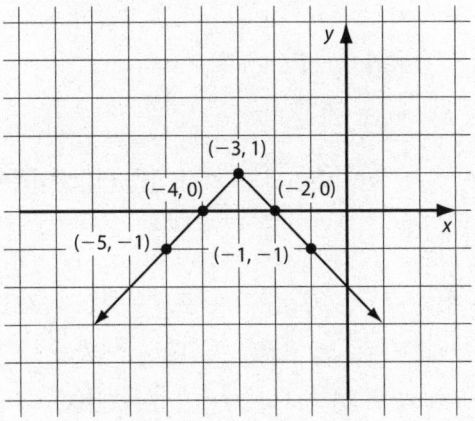

Example 8: Given the graph $y = |x + 2|$ shown here, if we rotated it 90° clockwise, what would the figure look like?

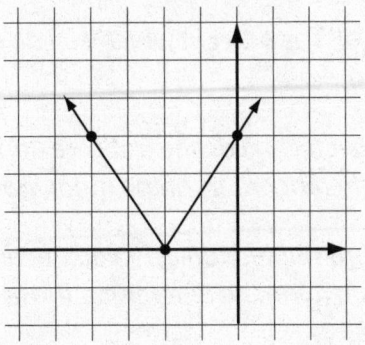

Solution: What you should do is turn the test booklet 90° clockwise to see the answer. The graph looks like this:

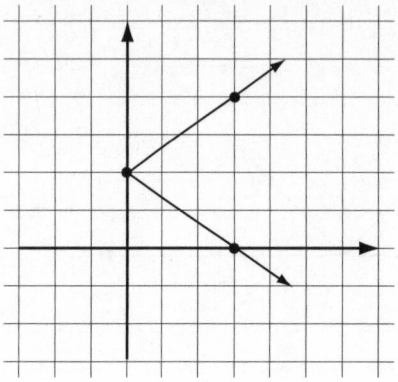

COMPOSITION

Suppose we have a function with domain D, range R_1, and map f. Also suppose we have a second function with domain R_1, range R_2, and map g. Is there a function that takes us directly from D to R_2? The answer is: Of course; otherwise, why would I write this?

The map is called the composition $g \circ f$ (read, "g circle f").
The definition is $g \circ f(x) = g(f(x))$.

Example 9: Suppose $f(x) = 5x + 7$ and $g(x) = x^2 + 10$. Find $g(f(4))$, $f(g(4))$, $g(f(x))$, and $f(g(x))$.

Solution: $g(f(4))$: $f(4) = 5(4) + 7 = 27$; $g(27) = 27^2 + 10 = 739$.
$f(g(4))$: $g(4) = 4^2 + 10 = 26$; $f(26) = 5(26) + 7 = 137$.
$g(f(x))$: We know $g(x) = x^2 + 10$; so $g(f(x)) = [f(x)]^2 + 10 = (5x + 7)^2 + 10 = 25x^2 + 70x + 59$.
$f(g(x))$: We know $f(x) = 5x + 7$; so $f(g(x)) = 5[g(x)] + 7 = 5(x^2 + 10) + 7 = 5x^2 + 57$.

Note *Maps are not commutative in general. The order in which you do a map matters; you get different answers with different orders.*

Note *If we were to draw pictures here, or get technical and indicate domains and ranges, the pictures, domains, and ranges for different orders of the maps would be very different.*

Note *If some of you eventually go into math-related topics, if you are doing research in math-related books pre-1970, this definition might be different in those books.*

INVERSE FUNCTIONS

Suppose we have a function map f, domain D, and range R. Do we have a function that takes R back into D. If D and R are in **1–1 correspondence** (can be paired), the answer is yes. It is called the **inverse map** f^{-1} $\left(\text{read, "}f\text{ inverse"; it does not mean } \frac{1}{f}\right)$.

Graphically, a function (which allows a vertical line to hit the graph only once) can have an inverse only if all horizontal lines can hit the graph no more than once. Let's look at an example.

Example 10: Let $f(x) = 5x + 7$, domain $D = \{1,5,9,20\}$. Find $f^{-1}(x)$, the inverse of $f(x)$.

Solution: First, find the range $R = \{f(1), f(5), f(9), f(20)\} = \{12, 32, 52, 107\}$. Notice that the domain and range are in $1-1$ correspondence.

To find the inverse, switch the domain and range. So $D = \{12, 32, 52, 107\}$.

To find the inverse map, we have to solve for x. Let $f(x) = y = 5x + 7$. So $5x = y - 7$ and $x = \frac{y-7}{5}$.

Now we have a notation change. The y is in the domain; we rename it x. The "x" we solved for is the new function f^{-1}. So $f^{-1}(x) = \frac{x-7}{5}$ with $D = \{12, 32, 52, 107\}$. Let's check it out.

$f^{-1}(12) = \frac{12-7}{5} = 1$; $f^{-1}(32) = \frac{32-7}{5} = 5$; $f^{-1}(52) = \frac{52-7}{5} = 9$; and $f^{-1}(107) = \frac{107-7}{5} = 20$.

Note *We have studied inverses in the past. The inverse of adding is subtracting; the inverse of subtracting is adding; the inverse of multiplying is dividing (excluding 0); and the inverse of dividing is multiplying. There are more. If we restrict the domain, the inverse of squaring (or any even root) is square rooting or any even "rooting." For odd powers, no restriction is necessary. The inverse of cubing is cube rooting, etc. Shortly, we will study another inverse. The inverse of exponentiation is logarithms.*

1 TO 1 FUNCTION (1–1 FUNCTION)

This topic is related to 1–1 correspondence. We will do this formally.

1–1 function: If $f(a) = f(b)$, then $a = b$.

Let's give examples of one function that is 1–1 and one that isn't.

Example 11: Let $f(x) = 2x$. Is $f(x)$ a 1–1 function?

Solution: If $f(a) = f(b)$, in symbols, $2a = 2b$, does $a = b$ always? The answer is yes since we can cancel the 2's. So $f(x) = 2x$ is a 1–1 function.

Example 12: Let $f(x) = x^2$. Is $f(x)$ a 1–1 function?

Solution: If $f(a) = f(b)$, in symbols $a^2 = b^2$, does $a = b$ always? The answer is no! $a = b$ or $a = -b$.

If we draw a picture of Example 11, it is a function that always increases. Curves that always increase or always decrease are 1–1 functions.

For Example 12, the curve both increases and decreases. It is *not* 1–1. By restricting the domain, we can make it 1–1. For example, we can allow only positive values for x.

Note *We have already used 1–1 without mentioning it. If $2^{3x+7} = 2^{5x-3}$, then $3x + 7 = 5x - 3$. Here $f(x) = 2^x$. If $f(a) = f(b)$, $2^a = 2^b$, and then $a = b$. If you draw 2^x, it is an increasing function.*

Example 13: Let $f(x) = x^2 + 5x + 9$, $g(x) = 4x + 2$, $F(x) = \dfrac{x}{x+3}$, $G(x) = \dfrac{1}{x}$. Find the values of the following functions:

Problem	Solution
a. $f(4)$	$4^2 + 5(4) + 9 = 45$
b. $g\left(-\frac{1}{2}\right)$	$4\left(-\frac{1}{2}\right) + 2 = 0$
c. $f(g(3))$	$f(14) = 275$
d. $g(f(1))$	$g(15) = 62$
e. $g(f(x))$	$g(x^2 + 5x + 9) = 4(x^2 + 5x + 9) + 2 = 4x^2 + 20x + 38$
f. $f(g(x))$	$f(4x + 2) = (4x + 2)^2 + 5(4x + 2) + 9 = 16x^2 + 36x + 23$
g. $F(G(x))$	$\dfrac{G(x)}{G(x) + 3} = \dfrac{\frac{1}{x}}{\frac{1}{x} + 3} \times \dfrac{x}{x} = \dfrac{1}{1 + 3x}$

h. $F^{-1}(x)$ First, find $F(x) = y = \dfrac{x}{x+3}$; so $(x+3)y = xy + 3y = x$;

$xy - x = x(y-1) = -3y$; so $x = \dfrac{-3y}{y-1}$ or $\dfrac{3y}{1-y}$.

We now need a notation change. $F^{-1}(x) = \dfrac{3x}{1-x}$.

i. $\dfrac{f(x+h) - f(x)}{h}$ $\dfrac{(x+h)^2 + 5(x+h) + 9 - (x^2 + 5x + 9)}{h} =$

$\dfrac{2xh + h^2 + 5h}{h} = 2x + h + 5$

The last question is a problem I hope you've seen. It is called the difference quotient. Some students study this in Algebra 2 or Precalc without knowing why. (You should always ask why or why it's true.) Here's the secret. SSSHHHHHH. It's at the beginning of calculus. SSSHHHHHH!

Let's go on to logarithms.

LOGARITHMS

Let us first define what is meant by logarithm. We write

$$\log_b x = y$$

(read as log(arithm) of x to the base b is y) if $b^y = x$. In words, $\log_{\text{base}}$ answer = exponent, because $\text{base}^{\text{exponent}}$ = answer.

Note *This is a very strange definition and must be practiced, especially if you haven't had much experience with it.*

Example 14: What is the value of $\log_3 81$?

Solution: $\log_3 81 = 4$ because $3^4 = 81$.

The **base** may be only positive numbers other than 1. There are two common bases. If we write "log," it means $\log_{10}$. The reason we use 10 as the base is because we have 10 fingers (really!). If we write "ln," it means log e, where e is a number like pi; $e \approx$ (approximately equals) 2.7. Although it doesn't stand for it, anything to do with e is easy!

The y values, the logarithms, can be any real number.

The x values must be positive because a positive number to any power **must** be positive!

Example 15: Find x:

Problem	Solution
a. $\log_5 125 = x$	$5^x = 125$, so $x = 3$
b. $\log_4 x = 3$	$x = 4^3 = 64$
c. $\log_x 81 = 2$	$x^2 = 81$, so $x = 9$ (only, because bases can't be negative)

Because logarithms are exponents, the laws of logarithms are the laws of exponents.

Law	Example
1. When you multiply, you add exponents; the same is true for logs. $\log_b cd = \log_b c + \log_b d$	$\log 6 = \log (2)(3) = \log 2 + \log 3$; $\ln dry = \ln d + \ln r + \ln y$
2. When you divide, you subtract exponents; the same is true for logs. $\log_b \frac{c}{d} = \log_b c - \log_b d$	$\log_7 \frac{3}{5} = \log_7 3 - \log_7 5$
3. A power to a power? You multiply exponents; the same is true for logs. $\log_b c^d = d \log_b c$	$\log 128 = \log 2^7 = 7 \log 2$
4. Any number to the first power is that number. $\log_b b = 1$, because $b^1 = b$.	$\log_3 3 = 1$, because $3^1 = 3$; $\log 10 = 1$ because $10^1 = 10$; $\ln e = 1$, because $e^1 = e$.
5. Any nonzero number to the zero power is 1. $\log_b 1 = 0$, because $b^0 = 1$, where b is any base.	$\log_2 1 = 0$, because $2^0 = 1$; $\ln 1 = 0$ because $e^0 = 1$
6. Log is a 1–1 function.	If $\log_b x = \log_b 3$, then $x = 3$.
7. Log is an increasing function.	$\log x < \log 6$, if $x < 6$.
8. $b^{\log_b x}$ is a really strange way to write x.	$9^{\log_9 5} = 5$.

9. $\log_b b^x = x$. $\log_7 7^5 = 5$.

10. $\log_c d = \dfrac{\log_b d}{\log_b c}$. $\log_3 7 = \dfrac{\log_{10} 7}{\log_{10} 3}$.

The last law needs some explanation. We do not know anything about base 3 (only about 10 and *e*). We can rewrite the problem as log 7 divided by log 3. This is *not*, *Not*, *NOT* $\log\left(\dfrac{7}{3}\right) = \log 7 - \log 3$.

In calculus, if you can do the next example and absolutely no other, it is more than 50 percent of what you need to know about logs.

Example 16: Write as simpler logs with no (or fewer) exponents: $\log_b \dfrac{p^4 q^9}{v^3\sqrt{u}}$

Solution: We add the logs that are products on the top, subtract the products on the bottom; exponents come down as coefficients: $4\log_b p + 9\log_b q - 3\log_b v - \dfrac{1}{2}\log_b u$.

Unfortunately, in Algebra 2 or Precalc, this type of problem is a very small part of the topic.

Here are two examples of problems you need to be able to solve.

Example 17: Solve for x: $\log_3 x + \log_3 (x + 6) = 3$

Solution: By the first law of logs, we get $\log_3 x(x + 6) = 3$. By the definition of logs, $x(x + 6) = 3^3$; $x^2 + 6x - 27 = (x + 9)(x - 3) = 0$. x can't be -9 because we can take the log of positive numbers only. The answer is $x = 3$. For fun, let's check: $\log_3 3 + \log_3 9 = 1 + \log_3 3^2 = 1 + 2\log_3 3 = 1 + 2(1) = 3$.

Example 18: Solve for x: $\log_4 x - \log_4 (x - 2) = 2$.

Solution: By the second law of logarithms, we get $\log_4 \dfrac{x}{x-2} = 2$. And by the definition of logs, $4^2 = \dfrac{16}{1} = \dfrac{x}{x-2}$. By cross-multiplying, we get $16x - 32 = x$; so $x = \dfrac{32}{15}$. As long as this answer does not give the log of a negative number (it doesn't), this is the answer. Otherwise, this problem would have no answer.

Example 19: Solve for x: $9^{4x+5} = 6^{2x+8}$.

Solution: Notice that 9 and 6 cannot be taken to the same base. The only way to get the x's off the exponents is to take logs of both sides, to get $(4x + 5)\log 9 = (2x + 8)\log 6$. So $x = \dfrac{8\log 6 - 5\log 9}{4\log 9 - 2\log 6}$.

Note *The algebra can be done in one step!!! Yes, it can. If we multiply out each side, getting all terms without x to the right, and all terms with x to the left, and factor out the x, the coefficient of x goes to the bottom of the fraction. You can do it if you try.*

Note *Let's do it in symbols. After multiplying the parentheses out, we get $Ax + B = Rx + S$. Rearranging, we get $Ax - Rx = (A - R)x = S - B$; so $x = \dfrac{S - B}{A - R}$.*

Note *On a test, I allow the students to leave this answer since pushing buttons on a calculator is silly. However, your teacher or the ACCUPLACER might ask you to do this. If you do, place a set of brackets around the top and a set of brackets around the bottom. On the calculator it is the top bracket divided by the bottom bracket.*

Note *Instead of using log, you could have used ln.*

Example 20: Find x.

Problem	Solution
a. $\log_{25} x = -\dfrac{3}{2}$	$\dfrac{1}{125}$
b. $\log_x 81 = 4$	3
c. $\log_{81} 27 = x$	$\dfrac{3}{4}$
d. $\log_7 7^2 = x$	2
e. $\log_9 1 = x$	0
f. $e^{\ln 7} = x$	7
g. $8^{2\log_8 5}$	25
h. $\log 5 + \log(2x + 3) = \log 85$	7
i. $\log_4\left(\dfrac{x-3}{2}\right) = 3$	131

j. $\log_2 x = 6 - \log_2 4x$ — 4

k. $4^{7x-5} = 3^{6x-8}$ — $\dfrac{5\log 4 - 8\log 3}{7\log 4 - 6\log 3}$

Let's try some more multiple-choice exercises.

Exercise 5: $\log_4 x = -2$. $x =$

A. -16
B. -8
C. $-\dfrac{1}{16}$
D. $\dfrac{1}{16}$
E. $\dfrac{1}{8}$

Exercise 6: $\log_{16} x = \dfrac{-3}{4}$. $x =$

A. 12
B. 8
C. $\dfrac{1}{8}$
D. $-\dfrac{1}{8}$
E. $-\dfrac{1}{12}$

Exercise 7: $\log_x 81 = 4$. $x =$

A. $\sqrt{3}$
B. 3
C. 9
D. $\dfrac{81}{4}$
E. There is more than one answer.

Exercise 8: $\log_{32} 64 = 5x$. $x =$

A. $\dfrac{1}{10}$
B. $\dfrac{6}{25}$
C. $\dfrac{2}{5}$
D. $\dfrac{1}{2}$
E. $\dfrac{6}{5}$

Exercise 9: $\log_2 x = 5 - \log_2 (x + 14)$. $x =$

A. 2 only
B. 2 and -16
C. -16 only
D. no real number
E. all real numbers

Let's look at the answers.

Answer 5: D: $4^{-2} = \left(\frac{1}{4}\right)^2 = \frac{1}{16}$.

Answer 6: C: $16^{-3/4} = \frac{1}{16^{3/4}} = \frac{1}{\left(\sqrt[4]{16}\right)^3} = \frac{1}{2^3} = \frac{1}{8}$.

Answer 7: B: $x^4 = 81$; $x = 3$; only positive answers are allowed.

Answer 8: B: $32^{5x} = 64$; $(2^5)^{5x} = 2^6$. When the bases are the same, the exponents must also be equal, so $25x = 6$; $x = \frac{6}{25}$.

Answer 9: A: This is an incredibly tricky problem until you see the trick. You must put both logs on the same side! $\log_2 x + \log_2 (x + 14) = \log_2 x(x + 14) = 5$; so $x(x + 14) = 2^5 = 32$. Then $x^2 + 14x - 32 = (x + 16)(x - 2) = 0$; so $x = 2$, and $x = -16$, which is not allowed. Every once in a while the ACCUPLACER gives choices for which you can see that only one can possibly be correct, even without doing the problem. From the expression $\log_2 x$ given in the exercise, we know that x must be greater than 0. Immediately, choices B, C, and E must be wrong. Substituting 2 in the problem, we see that it checks. If it didn't check, then the answer had to be no solution, the only one left.

Logarithms and their inverses, exponentiation (they are inverses, or opposites, similar to adding and subtracting), are the most interesting math-related topics below the level of calculus. Let me tell you the two stories I always tell at this point.

The first involves the Richter scale for earthquakes, which is roughly a logarithmic scale. Did you ever notice that a 3 causes a little shaking, a 5 causes minor damage, and at a 7, all #@$$#& breaks loose. On the Richter scale, to go between two numbers, the power of the earthquake is multiplied by 1,000! A Richter scale 7 is a million times as strong as a Richter scale 3. Hopefully, it won't happen, but an earthquake of magnitude 9 is supposed to hit California before 2030. If it does, the chances are that San Francisco will be gone, Oakland will be gone, the Golden Gate Bridge will be gone. Like in a Superman movie, Nevada might become the West Coast.

Another story: This is one of the few stories for which I tell the punch line first. This is when the little guy wins. In 1905, some guy—I don't remember his name—invented a photographic process of some kind and brought it to a large company. The company said it couldn't use it, and six years later came out with the product. The little guy took the big company to court for stealing his process. Now, normally the little guy loses automatically. Not this time!!! The little guy had taken a picture of San Francisco. As you may know, in 1906 a huge earthquake destroyed San Francisco, and the little guy had a picture of the pre-earthquake city. That is when the little guy wins—when all of San Francisco gets destroyed.

Okay, okay. Enough stories for now.

EXPONENTIAL GROWTH AND DECAY

This topic is perhaps more fascinating to discuss in a nonmathematical way. It relates to population growth. Assume humans came on the Earth 30,000 years ago (even though the actual figure is at least five times greater!). Count every human who ever lived from 30,000 years ago to the year 1900 (probably 1920 by now). There are more people alive today than that total number. Or go back to 1776. There were 4,000,000 people in the United States. Today there are 8,000,000 people in New York City alone. This shows the tremendous increase in population—exponential growth.

Let's do a little work on exponential growth and decay. We'll do it in two ways, the two ways that the ACCUPLACER might ask. It's still not the way I think it should be done, but it will be enough for now. The first way uses the exponential growth formula as shown in Examples 21 and 22.

The formula for continuous compound interest is given by $A = Pe^{rt}$, where P is the principal, r is the interest rate, t is the time, and A is the amount of money that you earn.

Example 21: Find A if we have \$2800 invested at 6% for 4 years.

Solution: $A = 2800\, e^{.06(4)}$. By pushing the proper buttons on the calculator, you would get the answer, which is \$3559.50.

Example 22: Find the interest if \$3000 becomes \$6000 in 14 years.

Solution: $6{,}000 = 3{,}000\, e^{14r}$; whenever the variable is in the exponent, it is necessary to isolate the term with the variable before taking logs (in this case ln). In this case we divide by 3,000 and get $2 = e^{14r}$. So $\ln 2 = 14r$ and $r = \frac{\ln 2}{14}$. This form of the answer is probably all you need, but if you use your calculator, you will see that this is .05, or 5%.

Now let's look at two examples of the second type of problem.

Example 23: Deb started with \$10,000. If her net worth triples every five years, how much will she have in 30 years?

Solution: In 5 years, the \$10,000 will be \$30,000; in 10 years, \$90,000; in 15 years, \$270,000; in 20 years, \$810,000; in 25 years, \$2,430,000; and in 30 years, \$7,290,000!

Example 24: A half-life is the time required for half of a radioactive substance to disintegrate. Suppose radioactive goo has a half-life of 15 minutes. If there are 240 pounds of radioactive goo, how much radioactive goo is left in an hour?

Solution: In 15 minutes, the 240 pounds will be 120 pounds; in 30 minutes, that 120 pounds will be 60 pounds; in 45 minutes, there will be 30 pounds; and in one hour, only 15 pounds will be left.

Note *I teach exponential growth different than these two methods. However, for this test, this is enough.*

Yesterday I came up with a problem the ACCUPLACER might actually ask. Here it is.

Example 25: Solve for x: $\ln(4x - 7) + 2 = 19$.

Solution: We isolate the log, and then use the definition of logs.
$\ln(4x - 7) = 17$; so $4x - 7 = e^{17}$; therefore $x = \dfrac{e^{17} + 7}{4}$.

Let's do the chapter test now.

Chapter 14 Test

For questions 1–6, let $f(x) = 5x + 3$, $g(x) = x^2 + 2x + 3$, $k(x) = \dfrac{3x - 5}{2x - 7}$, and $r(x) = \dfrac{1}{x}$.

1. $g(-3) =$
2. $f(g(2)) =$
3. $g(f(x)) =$
4. $k^{-1}(x) =$
5. $k(r(x)) =$
6. $\dfrac{g(x + h) - g(x)}{h} =$
7. Write as simpler logs with no exponents: $\ln\left(\dfrac{pq^6}{5r^7 \cdot \sqrt[3]{s}}\right)$.
8. Write as a single log: $\log 6 + \log 7 + 5\log x - \log e - \log 11$.
9. $3^{4\log_3 5} =$
10. $4\log_5 5^3 =$
11. $\log_7 1^{111} =$
12. If $\log 2 = .30$, $\log 3 = .48$ and $\log 5 = .70$, find $\log_2 3$.

For questions 13–20, solve for x.

13. $\log_x 64 = 6$.
14. $\log_{1/4} x = -\frac{3}{2}$.
15. $\log_{81} 27 = x$.
16. $\log_2 x + \log_2 2x = 5$.
17. $\log_5 x + \log_5 (x + 4) = \log_5 21$.
18. $\log_3 x - \log_3 (x - 3) = 3$.
19. $9^{3x+2} = 27^{4x-3}$.
20. $4 \log(2x - 6) + 7 = 75$.

For questions 21 and 22, find x; the answers can be in terms of logs, if necessary.

21. $4^{5x-6} = 7^{3x-2}$.
22. $4(10)^{3x+5} = 18$.

For questions 23–25: let $P = P_o(1.02)^t$, where P_o is the population of Earth and t is the number of years. Leave answers in terms of logs or exponentials. Suppose P_o is 6.5 billion people today.

23. How many people will there be in ten years? Do *not* use a calculator.
24. When will there be seven billion people? Answer in terms of ln.
25. In how many years will Earth's population double? Answer in terms of ln.

Answers to Chapter 14 Test

1. $(-3)^2 + 2(-3) + 3 = 6$.
2. $f(11) = 58$.
3. $(5x+3)^2 + 2(5x+3) + 3 = 25x^2 + 40x + 18$.
4. $k^{-1}(x) = y = \frac{3x - 5}{2x - 7}$; so $y(2x - 7) = 2xy - 7y = 3x - 5$; $2xy - 3x = x(2y - 3) = 7y - 5$; $x = \frac{7y - 5}{2y - 3}$; with the notation change, we get $k^{-1}(x) = \frac{7x - 5}{2x - 3}$.
5. $\dfrac{3\left(\frac{1}{x}\right) - 5}{2\left(\frac{1}{x}\right) - 7} = \dfrac{3 - 5x}{2 - 7x}$.
6. $\dfrac{(x + h)^2 + 2(x + h) + 3 - (x^2 + 2x + 3)}{h} = 2x + 2 + h$.

7. $\ln p + 6\ln q - \ln 5 - 7\ln r - \frac{1}{3}\ln s$.
8. $\log\left(\frac{42x^5}{11e}\right)$.
9. $5^4 = 625$.
10. $4(3) = 12$.
11. 0.
12. $\frac{\log 3}{\log 2} = \frac{.48}{.30} = 1.6$.
13. $x^6 = 64; x = 2$.
14. $\left(\frac{1}{4}\right)^{-\frac{3}{2}} = 4^{\frac{3}{2}} = 2^3 = 8$.
15. $81^x = 27; (3^4)^x = 3^3$; so $4x = 3$; and $x = \frac{3}{4}$.
16. $\log_2 x(2x) = 5; 2x^2 = 32; x = 4$ (–4 is no good since you can't have the log of a negative number).
17. $\log_5 x(x + 4) = \log_5 21$; so $x^2 + 4x = 21$ or $x^2 + 4x - 21 = (x + 7)(x - 3) = 0$ (again, only $x = 3$ because you can't have the log of a negative number).
18. $\log_3\left(\frac{x}{x-3}\right) = 3$; so $\frac{x}{x-3} = 3^3 = \frac{27}{1}$; cross-multiply to get $27(x - 3) = x; x = \frac{81}{26}$.
19. $(3^2)^{3x+2} = (3^3)^{4x-3}$; by 1 – 1, we get $2(3x + 2) = 3(4x - 3); x = \frac{13}{6}$.
20. $\log(2x - 6) = 17; 2x - 6 = 10^{17}; x = \frac{10^{17} + 6}{2}$.
21. $(5x - 6)\ln 4 = (3x - 2)\ln 7; x = \frac{6\ln 4 - 2\ln 7}{5\ln 4 - 3\ln 7}$; log is also OK.
22. $10^{3x+5} = 4.5; 3x + 5 = \log 4.5; x = \frac{\log 4.5 - 5}{3}$.
23. $P = 6.5 \times 10^9 \times (1.02)^{10}$.
24. $7 \times 10^9 = 6.5 \times 10^9 (1.02)^t; \frac{14}{13} = 1.02^t; t = \frac{\ln\left(\frac{14}{13}\right)}{\ln(1.02)}$.
25. $13 \times 10^9 = 6.5 \times 10^9 \times 1.02^t; 2 = 1.02^t$; so $t = \frac{\ln 2}{\ln 1.02}$. By calculator, Earth's population, assuming 2% growth annually, doubles every 35 years. Scary! So in 70 years, there will be 26 billion people, and in 105 years, there will be 52 billion of us!!

And now on to trig.

Answer to "Bob Asks": The 46th day. After 45 days, the monkey is 45 feet high. On the 46th morning, it climbs 5 feet to 50 feet, getting out of the hole.

CHAPTER 15: *All About Trig*

"How much dirt is there in a hole that is 5 feet by 4 feet by 9 inches?"

The secret of making trigonometry (trig) easy—yes, I mean easy—is to draw triangles. One topic, identities, is impossible to relate to triangles and is the only difficult topic in trig. It is treated very, very mildly in any placement test.

Let's start with angles. For trig purposes, we will have two measures of angles. The first one you know. There are 360 degrees in a circle. You may notice that there are many multiples of 60 in our measurements. More than 7,000 years ago, the Babylonians counted in 60s mainly because they thought there were 360 days in a year; 360 is a great number because it has many factors: 1, 2, 3, 4, 5, 6, 8, 9, 10, 12, 15, 18, 20, 24, 30, 36, 40, 45, 60, 72, 90, 120, 180, and 360. Whew!!!

Any intelligent life coming from another world would not have heard of degrees (they would quickly learn, of course!). However, all would have heard of **radians**.

Lay a radius on the circumference of a circle. The angle formed is said to be one radian.

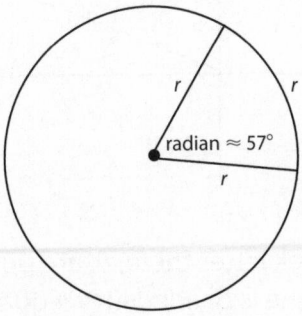

There are a little over 6, or exactly 2π radians in a circle. So $2\pi = 360°$. This means $1° = \frac{\pi}{180}$ radians, and a radian is $\approx 57°$. However, we usually express radians in terms of π, so we should be concerned with only two conversions:

1. To change from degrees to radians, multiply by $\frac{\pi}{180}$.
2. To change from radians to degrees, multiply by $\frac{180}{\pi}$.

Example 1: Change 45° to radians.

Solution: $45 \times \frac{\pi}{180} = \frac{\pi}{4}$ radians.

Example 2: Change from $\frac{\pi}{6}$ radians to degrees.

Solution: $\frac{\pi}{6} \times \frac{180}{\pi} = 30°$.

So we see that $\frac{\pi}{6}$ radians = 30°, and $\frac{\pi}{4}$ radians = 45°.

It also is a good idea to learn the most important multiples of 30° and 45°:

$\frac{\pi}{6} = 30°, \frac{\pi}{4} = 45°, \frac{\pi}{3} = 60°, \frac{\pi}{2} = 90°, \pi = 180°, \frac{3\pi}{2} = 270°$, and $2\pi = 360°$.

If you know more, that is better, but this is enough.

TRIG RATIOS

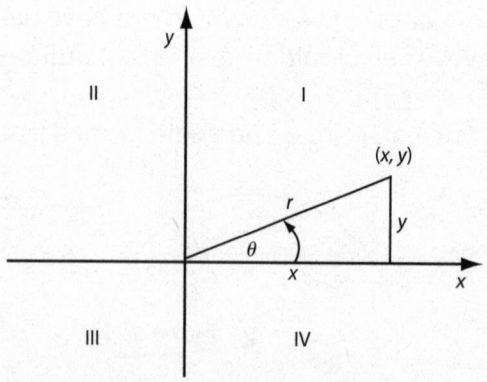

Angles, represented here by the Greek letter theta, θ, are positive if measured counterclockwise from the positive x-axis. The plane is divided into 4 quadrants, as indicated.

Locate the point (x, y) at a distance r from the origin, where $r = \sqrt{x^2 + y^2}$ is always positive.

Note *Always draw the triangle up or down to the x-axis.*

Referring to this triangle, which is a right triangle, we can define the six basic trigonometric ratios:

Function	Abbreviation	Ratio	Function	Abbreviation	Ratio
sine θ	$\sin\theta$	$\frac{y}{r}$	cotangent θ	$\cot\theta$	$\frac{x}{y}$
cosine θ	$\cos\theta$	$\frac{x}{r}$	secant θ	$\sec\theta$	$\frac{r}{x}$
tangent θ	$\tan\theta$	$\frac{y}{x}$	cosecant θ	$\csc\theta$	$\frac{r}{y}$

You must know these definitions!

Example 3: Let $\sin A = \frac{7}{10}$ in quadrant II. Find $\sec A$.

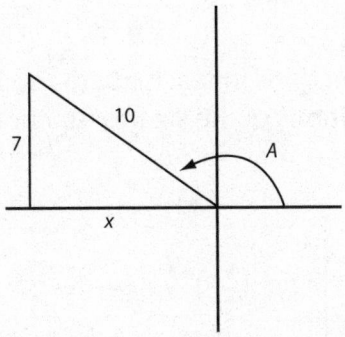

Solution: $\sin A = \frac{7}{10} = \frac{y}{r}$. Then, to find x, we can let $y = 7$ and $r = 10$, and we can use the Pythagorean Theorem: $x = \pm\sqrt{10^2 - 7^2} = \pm\sqrt{51}$. Because x is to the left, it is a negative x value. Then $\sec A = \frac{r}{x} = \frac{10}{-\sqrt{51}} = \frac{-10\sqrt{51}}{51}$.

Note *Once you get x, you have all six of the trig functions!*

Example 4: Let $\tan B = \frac{5}{9}$ in quadrant III. Find $\cos B$.

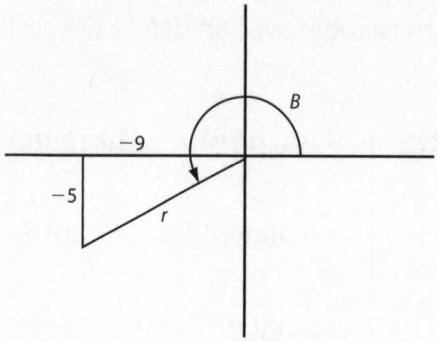

Solution: $\tan B = \frac{5}{9} = \frac{y}{x}$. In quadrant III, both x and y are negative. We let $y = -5$ and $x = -9$.

Then $r = \sqrt{9^2 + 5^2} = \sqrt{106}$, and $\cos B = \frac{x}{r} = \frac{-9}{\sqrt{106}} = \frac{-9\sqrt{106}}{106}$.

Note *Remember, r is always positive.*

It is necessary to know the signs of the trig functions for each quadrant. You should know the following sign diagram. If you know the definitions of the six trig functions, you should be able to figure this out yourself.

II $x < 0$ $y > 0$ $r > 0$ only sin, csc positive	I $x > 0$ $y > 0$ $r > 0$ all positive
$x < 0$ $y < 0$ $r > 0$ only tan, cot positive III	$x > 0$ $y < 0$ $r > 0$ only cos, sec positive IV

FINDING TRIG VALUES FOR MULTIPLES OF 30°, 45°, 60°, AND 90°

For this section, we have to recall the 45°-45°-90° and 30°-60°-90° triangles we discussed in Chapter 10. Let's give specific values for the sides of these triangles.

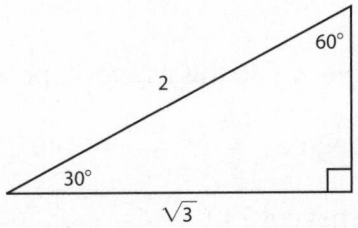

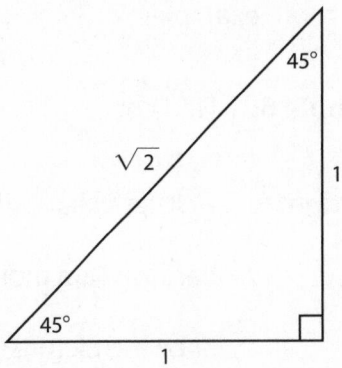

In the 30°-60°-90° triangle, if we let the side opposite the 30° angle = 1, then we know the hypotenuse is twice that, or 2, and we can find the side opposite the 60° angle, $\sqrt{3}$, by using the Pythagorean Theorem.

In the 45°-45°-90° triangle, if we let the sides opposite both 45° angles = 1, the Pythagorean Theorem tells us the hypotenuse, the side opposite the right angle = $\sqrt{2}$.

Example 5: Find cos 150°.

Solution:

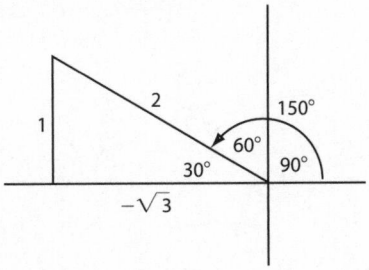

We draw the angle counterclockwise from the positive axis: 90° + 60° = 150°. The angle with the *x*-axis is 30°, giving us a 30°-60°-90° right triangle. The side opposite the 30° angle is 1, and it is +1 because it is "up." The side opposite the 60° angle is $-\sqrt{3}$. It is to the left, so it is negative. The hypotenuse is 2; as we have already seen, it is always positive. So $\cos 150° = \frac{x}{r} = \frac{-\sqrt{3}}{2}$.

Note *We always draw the triangle to the x-axis.*

Let's do two more examples.

Example 6: Find $\cot \frac{4\pi}{3}$.

Solution: First, let's change the angle to degrees. I do this wherever possible because I am more familiar with degrees. $\frac{4\pi}{3} \times \frac{180}{\pi} = 240°$. When we draw the picture (below), we see that $\cot 240° = \frac{x}{y} = \frac{-1}{-\sqrt{3}} \times \frac{\sqrt{3}}{\sqrt{3}} = \frac{\sqrt{3}}{3}$.

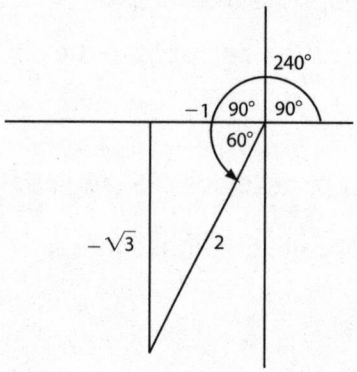

Note *The ACCUPLACER always rationalizes the denominator.*

Example 7: Find csc 315°.

Solution: Draw the figure. $\csc 315° = \frac{r}{y} = \frac{\sqrt{2}}{-1} = -\sqrt{2}$.

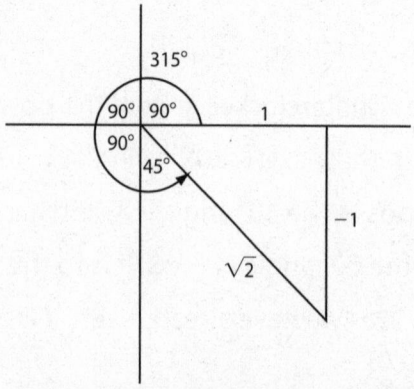

MULTIPLES OF 90°

To find the values of x, y, and r for angles on the axes, assume r is along the axis and equals 1. Then determine whether x and y are 0, +1, or −1 along that axis, and proceed as above.

Example 8: Find all six trig functions for 180°.

Solution:

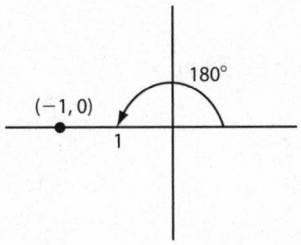

When we draw an angle of 180°, we wind up on the negative x-axis. Let $r = 1$, and then the point on the x-axis would be $(-1, 0)$ or $x = -1$ and $y = 0$. Once we know x, y, and r, we know all six trig functions. The ACCUPLACER would ask for only one, but let's do all six in this case.

$$\sin 180° = \frac{y}{r} = \frac{0}{1} = 0 \qquad \cos 180° = \frac{x}{r} = \frac{-1}{1} = -1$$

$$\tan 180° = \frac{y}{x} = \frac{0}{1} = 0 \qquad \cot 180° = \frac{x}{y} = \frac{-1}{0}, \text{ undefined}$$

$$\sec 180° = \frac{r}{x} = \frac{1}{-1} = -1 \qquad \csc 180° = \frac{r}{y} = \frac{1}{0}, \text{ undefined}$$

Note *For all multiples of 90°, two of the trig functions will always be 0, two will always be undefined (0 is the denominator), and two will both be either 1 or −1.*

Example 9:

Problem	Solution
a. Change 50° to radians	$\frac{5\pi}{18}$
b. Change $\frac{11\pi}{12}$ to degrees	165°

c. Change $7\pi°$ to radians	$\frac{7\pi^2}{180}$
d. $\cos x = \frac{11}{12}$ in quadrant IV; find $\sin x$	$-\frac{\sqrt{23}}{12}$
e. $\sin x = -\frac{3}{7}$ in quadrant III; find $\tan x$	$\frac{3\sqrt{10}}{20}$
f. $\tan x = -\frac{a}{b}$ in quadrant II; find $\sec x$	$-\frac{\sqrt{a^2 + b^2}}{b}$
g. $\cot 120°$	$-\frac{\sqrt{3}}{3}$
h. $\csc \frac{7\pi}{4}$	$-\sqrt{2}$
i. $\cos 0°$	1
j. $\tan \frac{\pi}{2}$	Undefined or none or ∞ (the denominator is 0)

SKETCHING SINES AND COSINES

You will not be asked to sketch sines and cosines, but some questions can be answered only if you know what the sketches look like. Here, we will do most everything in degrees, but we will refer to radians when appropriate.

All trig functions are **periodic**. That means they repeat. The **period** is the smallest interval in which the trig functions repeat. For sine and cosine (and secant and cosecant), the period is 360° or 2π. The period for tangent and cotangent is 180° or π.

If we were to graph $y = \sin x$ and $y = \cos x$, the pictures would look like these:

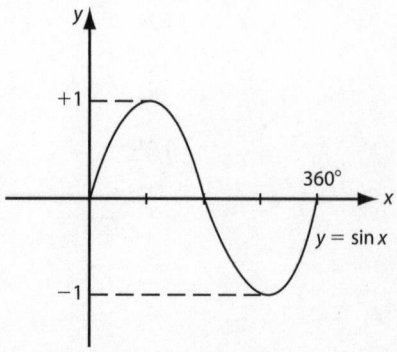

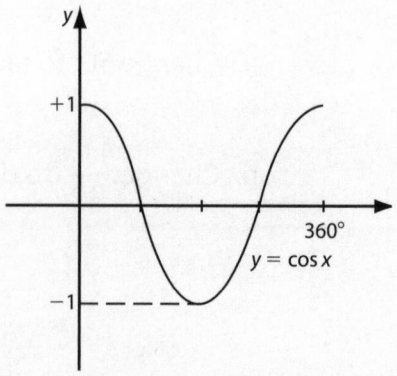

Example 10: Sketch $y = 4 \sin x$.

Solution:

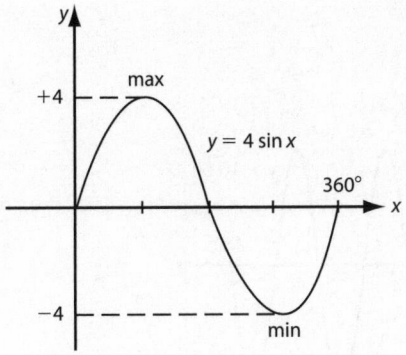

If you draw the figure, it goes four times as high as $y = \sin x$.

Note *The height is called the* ***amplitude****. The definition of amplitude is* $\frac{\text{max} - \text{min}}{2}$.

Example 11: Sketch $y = -6 \cos x$.

Solution:

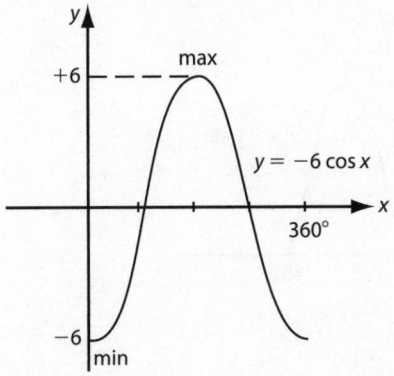

This graph is 6 times as high as $y = \cos x$. The minus sign means the curve is upside down. What is the amplitude? The amplitude is

$$\frac{\text{max} - \text{min}}{2} = \frac{6 - (-6)}{2} = \frac{12}{2} = 6.$$

Rule 1: If $y = A \sin x$ or $y = A \cos x$, the amplitude is $|A|$.

Example 12: Sketch $y = 10 \sin 4x$, in degrees.

Solution:

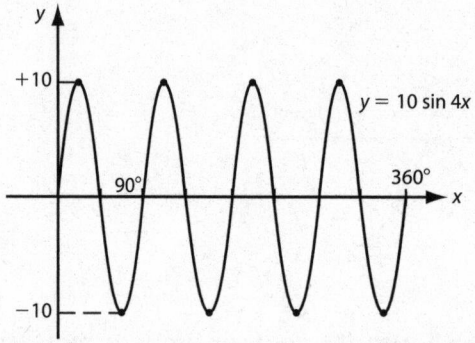

The amplitude is 10, and the graph is not upside down. The graph reaches 360° four times faster, so the period is $\frac{360°}{4} = 90°$.

Example 13: Sketch $y = -7 \sin \frac{x}{5}$.

Solution:

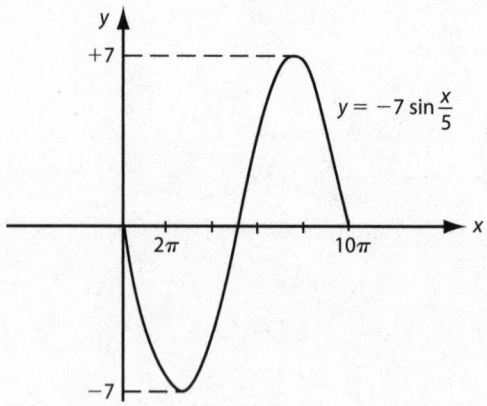

The amplitude is 7, and the graph is upside down. The period is $\frac{2\pi}{\frac{1}{5}} = 10\pi$, meaning the graph is "stretched out" five times more than $y = \sin x$.

So we can add to Rule 1:

Rules 1 and 2: If $y = A \sin Bx$ or $y = A \cos Bx$, the amplitude is $|A|$, and the period is $\frac{360°}{B}$, or $\frac{2\pi}{B}$ radians.

Example 14: Sketch $y = -2 \sin (3x - 120°)$.

Solution:

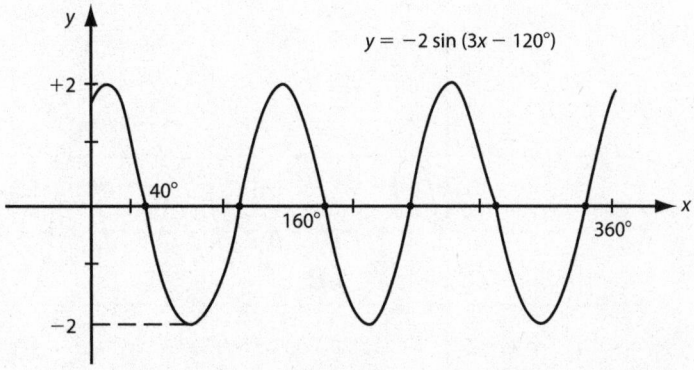

The amplitude is 2, the graph is upside down, and the period is $\frac{360°}{3} = 120°$. We know $\sin 0° = 0$, so we set $3x - 120° = 0°$, and we get $x = 40°$. This means a **left-right shift** of $y = 0$ is 40° to the right.

Example 15: Sketch $y = 9 \cos (6x + \frac{\pi}{5})$.

Solution:

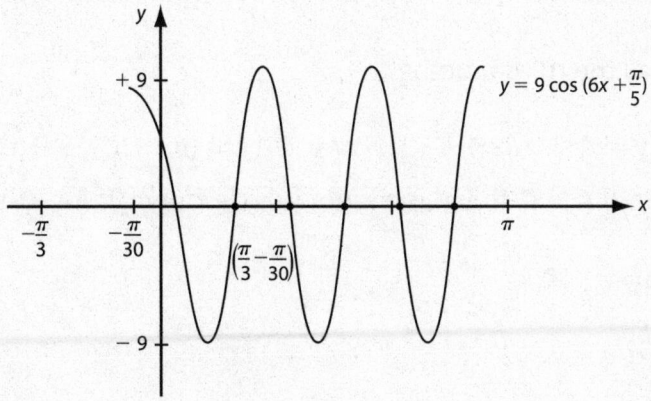

The amplitude is 9, the graph is not upside down, and the period is $\frac{2\pi}{6} = \frac{\pi}{3}$. We know $\cos 0°$ is at its maximum, so we set $6x + \frac{\pi}{5} = 0$, $x = -\frac{\pi}{30}$, so there is a $\frac{\pi}{30}$ shift of the maximum to the left.

So we can add another rule:

Rules 1, 2, and 3: If $y = A \sin(Bx + C)$ or $y = A \cos(Bx + C)$, the amplitude is $|A|$, the period is $\frac{360°}{B}$, or $\frac{2\pi}{B}$, and the left-right shift is $-\frac{C}{B}$.

Example 16: Sketch $y = 2 \sin(5x - 30°) + 7$.

Solution:

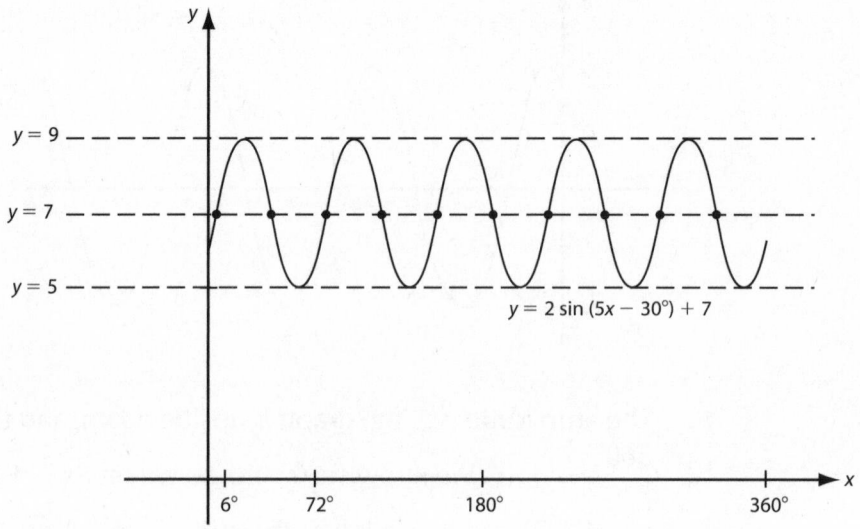

The amplitude is 2, the graph is not upside down, the period is $\frac{360°}{5} = 72°$, the left-right shift is $-\frac{(-30°)}{5} = 6°$ to the right, and the up-down shift is 7 up (sounds like a good name for a soft drink!).

So finally, we can put all the rules together:

Rules 1, 2, 3, and 4: If $y = A \sin(Bx + C) + D$ or $y = A \cos(Bx + C) + D$, then

- $|A|$ is the amplitude. If $A > 0$, the graph is not upside-down; if $A < 0$, the graph is upside down.
- Period is $\frac{360°}{B}$ or $\frac{2\pi}{B}$.
- Left-right shift is $-\frac{C}{B}$.
- Up-down shift is D: if $D > 0$, it is up; if $D < 0$, it is down.

That's it for sketching sines and cosines.

INVERSE TRIG FUNCTIONS

There is no perfect place to put this subject, so let's put it here.

First we need a brief review, as I always do. To have an inverse trig function, we need an inverse function. Before we have an inverse function we need a function. In order to have a function, a graph must pass the vertical line test. To have an inverse function, a graph must also past the horizontal line test. We just drew sine curves, which always pass the vertical line test. However, if you tried a horizontal line test, each line would hit the graph an infinite number of times. So we must restrict the original angles that the trig functions can take on. Also remember that in inverses, the range and domain of the original functions switch.

The *notation* for the inverse of sine is $\sin^{-1}A$ or arc sin A, and it means "the angle whose sine is A."

Typically, you should know the range for the inverses of sine, cosine, and tangent. Here is a list of their domains and ranges.

Function	Domain	Range
$\theta = \sin^{-1}A$	$-1 \le A \le 1$	$-\frac{\pi}{2} \le \theta \le \frac{\pi}{2}$, or $-90° \le \theta \le 90°$
$\theta = \cos^{-1}A$	$-1 \le A \le 1$	$0 \le \theta \le \pi$, or $0° \le \theta \le 180°$
$\theta = \tan^{-1}A$	$-\infty < A < \infty$	$-\frac{\pi}{2} \le \theta \le \frac{\pi}{2}$, or $-90° \le \theta \le 90°$

Inverse problems have only one answer. The positive inverse functions are all in quadrant I. Inverse cosine is in quadrant II, and inverse sine and inverse tangent are in quadrant IV.

The secret (shhhh) of doing inverse trig problems is to read it properly. The answer is an angle.

So reading $\sin^{-1}\left(\frac{1}{2}\right)$ as "The angle whose sine is $\frac{1}{2}$," we can draw the triangle in quadrant I. We see that the angle is 30°, or $\frac{\pi}{6}$.

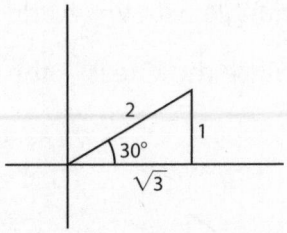

Note *The answer should always be in radians, at least when I taught it. However, we understand degrees better, so I put both!*

Example 17: $\sin^{-1}\left(-\frac{1}{2}\right) =$

Solution: This is read as, "Find the angle whose sine is minus one-half." The angle is in quadrant IV. Drawing the triangle, we see the answer is $-30°$, or $\left(-\frac{\pi}{6}\right)$.

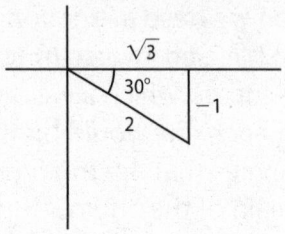

Note *Important! The answer is not 330°, or $\frac{11\pi}{6}$, since that angle is not in our acceptable range of $-90° \leq \theta \leq 90°$.*

Example 18: Find $\tan\left(\cos^{-1}\left(\frac{5}{8}\right)\right)$.

Solution: This is read as, "Find the tangent of the angle whose cosine is $\frac{5}{8}$." You can almost see the picture in your mind; $x = 5$ and $r = 8$; so $y = \sqrt{8^2 - 5^2} = \sqrt{39}$. Tangent $\theta = \frac{y}{x} = \frac{\sqrt{39}}{5}$.

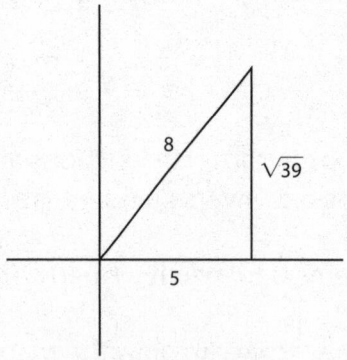

Example 19: Find $\sin\left(\tan^{-1}\left(\frac{q}{s}\right)\right)$.

Solution: We draw the triangle with $y = q$ and $x = s$. The hypotenuse is $r = \sqrt{q^2 + s^2}$. Since the sine is $\frac{y}{r}$, the answer is $\frac{q}{\sqrt{q^2 + s^2}}$.

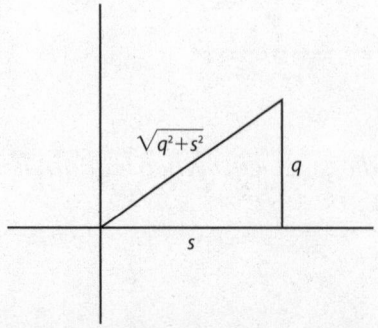

TRIGONOMETRIC IDENTITIES

An **identity** is an equation that is always true as long as it is defined. For example, $2x + 3x = 5x$ is an identity because no matter what number is used for x, $2x + 3x = 5x$.

You should know some trigonometric identities. Let's rewrite the basic six ratios again:

$$\sin\theta = \frac{y}{r} \qquad \cos\theta = \frac{x}{r} \qquad \tan\theta = \frac{y}{x}$$

$$\cot\theta = \frac{x}{y} \qquad \sec\theta = \frac{r}{x} \qquad \csc\theta = \frac{r}{y}$$

Now let's list the eight identities you absolutely, positively should know perfectly!

Identity	Reason
1. $\sin A \times \csc A = 1$	$\left(\frac{y}{r}\right)\left(\frac{r}{y}\right) = 1$
2. $\cos A \times \sec A = 1$	$\left(\frac{x}{r}\right)\left(\frac{r}{x}\right) = 1$
3. $\tan A \times \cot A = 1$	$\left(\frac{y}{x}\right)\left(\frac{x}{y}\right) = 1$
4. $\tan A = \frac{\sin A}{\cos A}$	$\frac{y}{r} \div \frac{x}{r} = \frac{y}{r} \times \frac{r}{x} = \frac{y}{x}$
5. $\cot A = \frac{\cos A}{\sin A}$	$\frac{x}{r} \div \frac{y}{r} = \frac{x}{r} \times \frac{r}{y} = \frac{x}{y}$
6. $\sin^2 A + \cos^2 A = 1$	Divide each term of $x^2 + y^2 = r^2$ by r^2.
7. $1 + \tan^2 A = \sec^2 A$	Divide each term of $x^2 + y^2 = r^2$ by x^2.
8. $1 + \cot^2 A = \csc^2 A$	Divide each term of $x^2 + y^2 = r^2$ by y^2.

Q Let's do a few multiple-choice exercises.

Exercise 1: $\sin G \times \cot G =$

A. $\sin G$

B. $\cos G$

C. $\tan G$

D. $\cot G$

E. $\sec G$

Exercise 2: Given $\sin(A + B) = \sin A \cos B + \cos A \sin B$, and given $\frac{5\pi}{12} = \frac{\pi}{4} + \frac{\pi}{6}$, then $\sin\left(\frac{5\pi}{12}\right) =$

A. $\frac{1}{2}$

B. $\frac{\sqrt{3}}{2}$

C. $\frac{\sqrt{6} - \sqrt{2}}{4}$

D. $\frac{\sqrt{6} + \sqrt{2}}{4}$

E. 1

A Let's look at the answers.

Answer 1: **B:** $\cot G = \frac{\cos G}{\sin G}$, so $\sin G \times \frac{\cos G}{\sin G} = \cos G$.

Answer 2: **D:** First change each of the radians to degrees. We get $75° = 45° + 30°$. So $\sin 75° = \sin 45° \cos 30° + \cos 45° \sin 30°$. Then $\sin(45° + 30°) =$

$$\frac{1}{\sqrt{2}} \times \frac{\sqrt{3}}{2} + \frac{1}{\sqrt{2}} \times \frac{1}{2} = \frac{\sqrt{3} + 1}{2\sqrt{2}} \times \frac{\sqrt{2}}{\sqrt{2}} = \frac{\sqrt{6} + \sqrt{2}}{4}.$$

You can get the same answer without changing radians to degrees, if you are comfortable working with radians.

TRIGONOMETRIC EQUATIONS

Trig equations are like regular equations except we must find the angles. The following three examples will give answers in both degrees and radians, with all answers between 0° and 360° (0 and 2π) including 0°.

Example 20: Solve for x: $\cos x + 2 \sin x \cos x = 0$.

Solution: Take out the common factor: $\cos x\,(2 \sin x + 1) = 0$. This means either $\cos x = 0$ or $\sin x = -\frac{1}{2}$. We know sine is negative in quadrants III and IV. We then draw the appropriate triangles and we see that $x = 210°$ and $330°$, or in radians, $\frac{7\pi}{6}$ and $\frac{11\pi}{6}$.

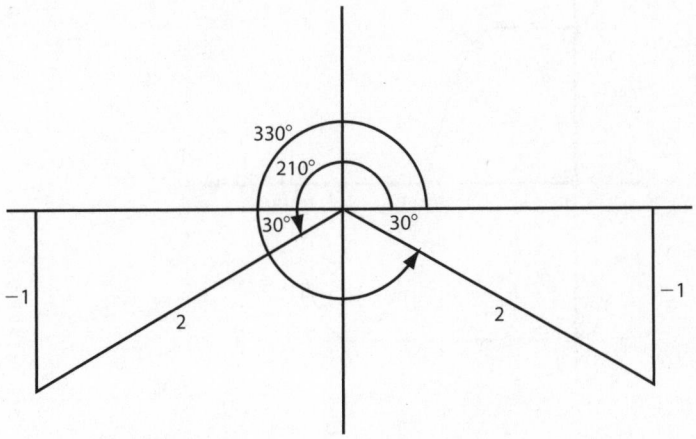

A little trick: Draw the $y = \cos x$ curve. From this curve, we see that $\cos x = 0$ at 90° or 270°, or in radians, $\frac{\pi}{2}$ or $\frac{3\pi}{2}$. So the answers to Example 20 are $x = 90°, 120°, 270°$, and $330°$, or $x = \frac{\pi}{2}$, $\frac{7\pi}{6}$, $\frac{3\pi}{2}$, and $\frac{11\pi}{6}$ radians.

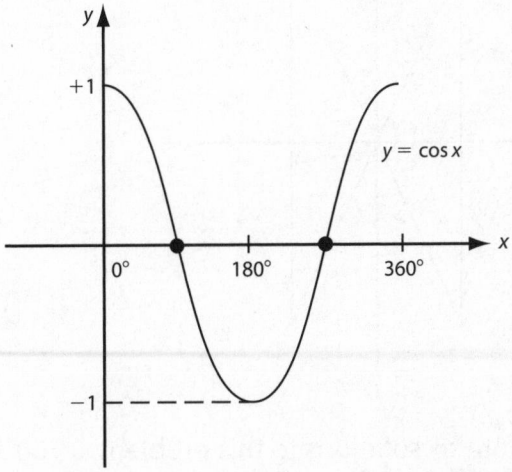

Note *Also note that $\cos 0° = \cos 360° = 1$, and $\cos 180° = -1$.*

Example 21: Solve for x: $4 \sin^3 x - 3 \sin x = 0$.

Solution: We get $\sin x\,(4 \sin^2 x - 3) = 0$. Thus, $\sin x = 0$ or $4 \sin^2 x = 3$; $\sin^2 x = \frac{3}{4}$; $\sin x = \pm \frac{\sqrt{3}}{2}$. $\sin x = 0$ for $x = 0°$ and $180°$ (see the sine curve below), or 0 and π radians. We don't include $360°$ or 2π because it is the same as 0.

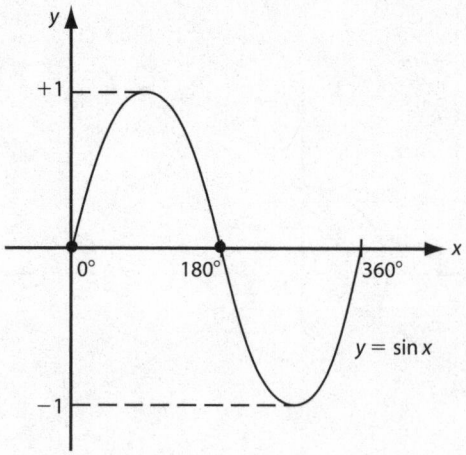

Next, we see $\sin x = +\frac{\sqrt{3}}{2}$ in quadrants I and II, and $x = -\frac{\sqrt{3}}{2}$ in quadrants III and IV. From the graph below, we can see that $x = 60°, 120°, 240°$, and $300°$, or in radians, $x = \frac{\pi}{3}, \frac{2\pi}{3}, \frac{4\pi}{3}$, and $\frac{5\pi}{3}$.

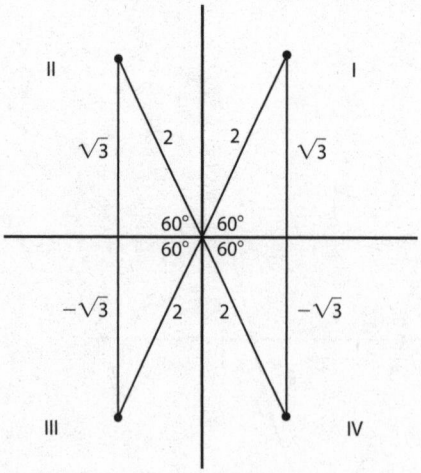

There are six solutions to this problem: 0°, 60°, 120°, 180°, 240°, and 300°, or 0, $\frac{\pi}{3}, \frac{2\pi}{3}, \pi, \frac{4\pi}{3}$, and $\frac{5\pi}{3}$ radians.

Note *$\sin 90°$ (or $\sin \frac{\pi}{2}$) $= 1$; $\sin 270°$ (or $\sin \frac{3\pi}{2}$) $= -1$.*

Example 22: Solve for x: $\tan^2 x + \sec x - 1 = 0$.

Solution: We let $\tan^2 x = \sec^2 x - 1$. The equation becomes $\sec^2 x - 1 + \sec x - 1 = \sec^2 x + \sec x - 2 = (\sec x + 2)(\sec x - 1) = 0$. So the roots of the equation are $\sec x = -2$ or $\sec x = 1$. But $\cos x = \frac{1}{\sec x}$, so we can rewrite the roots as $\frac{1}{\cos x} = -2$, or $\cos x = -\frac{1}{2}$ and $\cos x = 1$. From the graph of $y = \cos x$, we see that $\cos x = 1$ means $x = 0°$, or 0 radians. For $\cos x = -\frac{1}{2}$, $\cos x$ is negative in quadrants II and III, as seen in the graph below. Then $x = 120°$ and $240°$, or $\frac{2\pi}{3}$ and $\frac{4\pi}{3}$ radians.

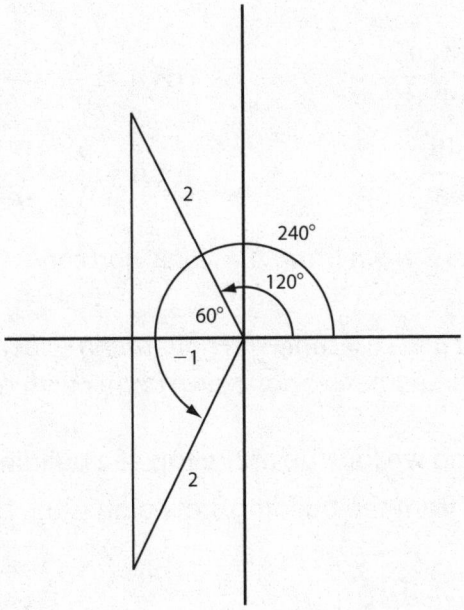

The answers to Example 22 are 0°, 120°, and 240°, or 0, $\frac{2\pi}{3}$, and $\frac{4\pi}{3}$ radians.

RIGHT-ANGLE TRIGONOMETRY

I believe that if you do non-right-angle trig first and then do right-angle trig last, right-angle trig seems very easy. If you do right-angle trig first, then non-right-angle trig is not so easy. That is why I'm following this order.

Take the definitions we have been using. Put the triangle in the first quadrant. Then remove the triangle from the x and y axes. You get the following right triangle:

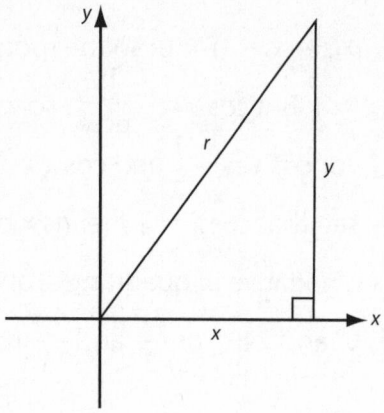

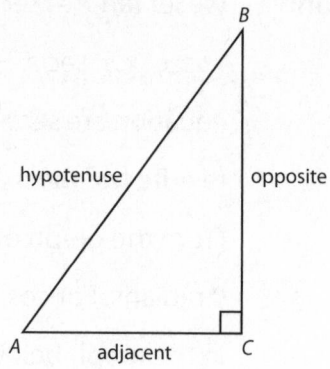

Notice: $\sin A = \dfrac{y}{r} = \dfrac{BC}{AB} = \dfrac{\text{opposite}}{\text{hypotenuse}}$ $\qquad$ $\cos A = \dfrac{x}{r} = \dfrac{AC}{AB} = \dfrac{\text{adjacent}}{\text{hypotenuse}}$

$\tan A = \dfrac{y}{x} = \dfrac{BC}{AC} = \dfrac{\text{opposite}}{\text{adjacent}}$ $\qquad$ $\cot A = \dfrac{x}{y} = \dfrac{AC}{BC} = \dfrac{\text{adjacent}}{\text{opposite}}$

Note *Notice that $\sin A = \cos B$, $\cos A = \sin B$, $\tan A = \cot B$, and $\cot A = \tan B$.*

Note *We don't bother with secant and cosecant with right triangles because no one does in the Age of the Calculator. Sometimes we don't even bother with cotangent.*

Example 23: From the ground we look up at the top of a building at an angle of 53°, if we are 9 feet from the building, how high is the building?

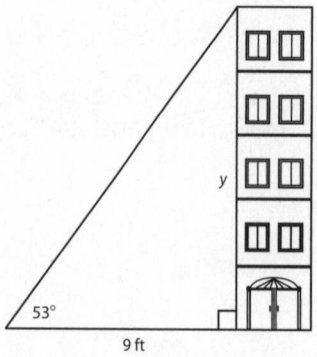

Solution: From the picture, we get that $\tan 53° = \dfrac{y}{9}$; $y = 9 \tan 53°$. $\tan 53° \approx 1.33$, so $y \approx 12$ feet.

Note *This angle is called the **angle of elevation.***

Example 24: From the top of a lighthouse 360 feet high, the **angle of depression** to a boat is 37°. How far is the boat from the bottom of the lighthouse, and how far from you is the boat?

Note *The **angle of depression** is the angle between a line parallel to the horizon and the object.*

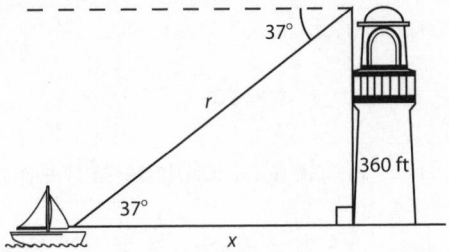

Solution: From the picture, by alternate interior angles (which are equal), the angle of depression equals the angle of elevation. So we have

$\tan 37° = \dfrac{360}{x}$; so $x = \dfrac{360}{\tan 37°} \approx 480$ feet, since $\tan 37° \approx 0.75$.

Now, $\sin 37° = \dfrac{360}{r}$, so $r = \dfrac{360}{\sin 37°} \approx 600$ feet, since $\sin 37° \approx 0.60$.

Note *If you have two sides missing in the same problem, always use a trig function that involves the side you are given, for three reasons: it is exactly correct (it is given); you may have made a mistake in the first part; and even if you are correct, the new side part you obtained is approximately correct.*

Example 25: In a 5–12–13 right triangle, find all the angles.

Solution: We know one angle is 90°! The tangent of the smallest angle is given by $\tan x = \dfrac{5}{12}$. So $x = \tan^{-1}\left(\dfrac{5}{12}\right)$. By calculator, this is approximately 23°. The third angle is then equal to 90° − 23°, or approximately 67°.

LAW OF SINES AND LAW OF COSINES

In geometry, we learned that the largest side of a triangle lies opposite the largest angle (see triangle *ABC*).

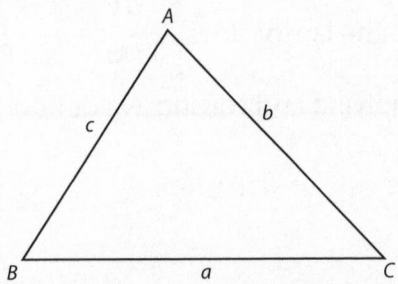

In fact, for any triangle (not just right triangles), the ratios of any side to the sine of the angle opposite that side are always the same. This is represented mathematically by the **law of sines**:

$$\frac{a}{\sin A} = \frac{b}{\sin B} = \frac{c}{\sin C}$$

The **law of cosines** states that for any triangle *ABC*:

$$c^2 = a^2 + b^2 - 2ab \cos C$$

In general, (side 1)^2 = (side 2)^2 + (side 3)^2 − 2(side 2)(side3)(cos of the angle opposite side 1).

Example 26: In a triangle *ABC*, let $c = 5$, $b = 6$ and $\angle A = 52°$. Find the length of side *a*.

Solution: We do not have a side and an angle opposite, so we need the law of cosines. We have $a^2 = b^2 + c^2 - 2bc \cos A$, or $a^2 = 6^2 + 5^2 - 2(6)(5) \cos 52°$. Then $a = \sqrt{61 - 60 \cos\ 52°}$, which we would do by calculator.

Note *Once you get four parts of a triangle, you could always use the law of sines for the other parts. If you know two angles, to get the third angle you would subtract the sum of these two from 180°.*

Example 27: In a triangle *DEF*, let $d = 7$, $e = 8$, and $f = 9$. Find the middle angle, *E*.

Solution: We have no angles, so we must use the law of cosines, which can be written as $e^2 = f^2 + d^2 - 2fd \cos E$, or $8^2 = 9^2 + 7^2 - 2(9)(7) \cos E$; $64 = 130 - 126 \cos E$; $\cos E = \frac{66}{126}$; $E = \cos^{-1}\left(\frac{11}{21}\right)$.

Note *E is the middle angle because e is the middle side (in size).*

Example 28: In a triangle *MNP*, let $\angle M = 40°$, $\angle N = 85°$, and $m = 10$ meters. Find side *p*.

Solution: $\angle P = 180° - (40° + 85°) = 55°$. We have a side and an angle opposite. We can use the law of sines, $\frac{m}{\sin M} = \frac{p}{\sin P}$, or $\frac{10}{\sin\ 40°} = \frac{p}{\sin\ 55°}$. By cross-multiplying and solving, we get $p = 10\left(\frac{\sin\ 55°}{\sin\ 40°}\right)$.

Chapter 15 Test

1. Csc 315° =
2. Tan $\frac{7\pi}{6}$ =
3. Cot 180° =

For questions 4–7, let $y = -10 \sin\left(6x - \frac{\pi}{3}\right) - 30$.

4. What is the amplitude?
5. What is the period?
6. What is the left-right shift? Indicate if it is left or right.
7. What is the up-down shift?

Questions 8–10 refer to a triangle with sides 3, 5, and 7.

8. Find the largest angle.
9. Find the middle angle.
10. Find the smallest angle.

11. A person looks up at the top of a 100-foot tree at an angle of 53°. How far from the tree is the person?
12. $\text{Cos}^{-1}\left(\frac{-1}{\sqrt{2}}\right) =$
13. $\text{Tan}\left(\sin^{-1}\left(\frac{2}{3}\right)\right) =$
14. If m and n are in the quadrant I, $\csc\left(\cot^{-1}\left(\frac{m}{n}\right)\right) =$
15. Solve for x, $0 \le x < 360°$ (also write angles in radians): $4\cos^3 x - \cos x = 0$.
16. Solve for x, $0 \le x < 360°$ (also write angles in radians): $\sqrt{3}\tan^2 x + \tan x = 0$.
17. $\text{Sin}\,(A + B) = \sin A \cos B + \cos A \sin B$. Find sin 105°.
18. If $\cos\frac{A}{2} = \pm\sqrt{\frac{1 + \cos A}{2}}$, find cos 165°.
19. If $\cos 2A = \cos^2 A - \sin^2 A$, find $\cos 2A$ if $\cos A = \frac{4}{b}$, where A is in quadrant I.
20. Show that $\sec^2 A + \csc^2 A = \sec^2 A \times \csc^2 A$.

Answers to Chapter 15 Test

1. $-\sqrt{2}$.
2. $\frac{\sqrt{3}}{3}$.
3. Undefined.
4. 10.
5. $\frac{2\pi}{6} = \frac{\pi}{3}$.
6. $6x - \frac{\pi}{6} = 0$; $6x = \frac{\pi}{3}$; $x = \frac{\pi}{3} \times \frac{1}{6} = \frac{\pi}{18}$, right shift.
7. Down 30.
8. $7^2 = 3^2 + 5^2 - 2(3)(5)\cos x$; $49 = 34 - 30\cos x$; $15 = -30\cos x$; $\cos x = -\frac{1}{2}$; $x = 120°$.
9. $\frac{7}{\sin 120°} = \frac{5}{\sin y}$; after cross-multiplying, $\sin y = \frac{5 \times \sin 120°}{7}$;
   $y = \sin^{-1}\left(\frac{5 \times \sin 120°}{7}\right)$; by calculator, $y = 38°$.
10. $180° - 120° - 38° = 22°$.
11. $\text{Tan } 53° = \frac{100}{x}$; $x = \frac{100}{\tan 53°} \approx 75$ feet.
12. If the angle is in the second quadrant; 135°, or $\frac{3\pi}{4}$. If the angle is in the third quadrant; 225°, or $\frac{5\pi}{4}$.
13. Let $y = 2$, $r = 3$; then $x = \sqrt{3^2 - 2^2} = \sqrt{5}$; $\tan = \frac{y}{x} = \frac{2}{\sqrt{5}} = \frac{2\sqrt{5}}{5}$.
14. We let $x = m$ and $y = n$; so $r = \sqrt{m^2 + n^2}$; cosecant $= \frac{r}{y} = \frac{\sqrt{m^2 + n^2}}{n}$.
15. By factoring, we get $\cos x(2\cos x - 1)(2\cos x + 1) = 0$. $\text{Cos } x = 0$ at 90° and 270° $\left(\text{or } \frac{\pi}{2} \text{ and } \frac{3\pi}{2} \text{ radians}\right)$ from the graph of $y = \cos x$. $\text{Cos } x = \frac{1}{2}$ in quadrants I and IV, so $x = 60°$ and 300° $\left(\text{or } \frac{\pi}{3} \text{ and } \frac{5\pi}{3} \text{ radians}\right)$. $\text{Cos } x = -\frac{1}{2}$ in quadrants II and III; so $x = 120°$ and 240° $\left(\text{or } \frac{2\pi}{3} \text{ and } \frac{4\pi}{3} \text{ radians}\right)$.
16. By factoring, we get $\tan x\left(\sqrt{3}\tan x + 1\right)$. $\text{Tan } x = 0$ if $x = 0°$ or 180° (or 0 or π radians) (note that 360° is not included). $\text{Tan } x = -\frac{1}{\sqrt{3}}$ in quadrants II and IV; so $x = 150°$ and 330°, $\left(\text{or } \frac{5\pi}{6} \text{ and } \frac{11\pi}{6} \text{ radians}\right)$.

17. Sin 105° = sin (60° + 45°) = sin (60°) cos (45°) + cos 60° sin (45°) = $\frac{\sqrt{6}+\sqrt{2}}{4}$.

18. Cos 165° = $\pm\sqrt{\frac{1+\cos 330°}{2}}$; since 165° is in the second quadrant, the cosine is negative, so we use the minus sign. The final (ugly) answer is $-\sqrt{\frac{1+\frac{\sqrt{3}}{2}}{2}}$, or $\frac{-\sqrt{2+\sqrt{3}}}{2}$.

19. Since $\sin^2 A + \cos^2 A = 1$ and A is in the first quadrant, $\sin^2 A = 1 - \cos^2 A = 1 - \left(\frac{4}{b}\right)^2 = \frac{b^2-16}{b^2}$; so $\cos 2A = \cos^2 A - \sin^2 A = \frac{16}{b^2} - \frac{b^2-16}{b^2} = \frac{32-b^2}{b^2}$.

20. Working only with the left side, we get
$\frac{1}{\cos^2 A} + \frac{1}{\sin^2 A} = \frac{\cos^2 A + \sin^2 A}{\cos^2 A \times \sin^2 A} = \frac{1}{\cos^2 A \times \sin^2 x} = \frac{1}{\cos^2 x} \times \frac{1}{\sin^2 x} =$
$\sec^2 A \times \csc^2 A$. It's a really weird identity; you have two trig functions; whether you add them or multiply them, you get the same answers, always, for any angle!

That's all for the trig. Let's get to the chapter on the topics that didn't seem to fit well into any other chapter.

Answer to "Bob Asks": There is no dirt in a hole.

CHAPTER 16: *Topics That Don't Fit in Anywhere Else*

"An ice cream cake has a radius of 7 inches and costs $10.00. If the radius is doubled, how much should the cake cost if the owner charges the same rate? (The height is the same.)"

The title of this chapter is not quite accurate. It is a chapter that contains material for which, if I wanted to place my students properly, not a single question would be asked. It's not that the math in this chapter is unimportant or unnecessary. The fact is that the placement exam is not 79 hours long, and certainly your school is not asking you to answer a 457-question test. In fact, if you are probably taking the standard ACCUPLACER test, you are answering exactly 49 questions: 17 arithmetic questions, 12 elementary algebra questions, and 20 college-level math questions, all multiple choice. For me, multiple choice doesn't show how well you know a question, but when you are testing many students quickly, it is the fastest way. If I have only 49 questions to find out how well you know your math, there are specific skills I want to see. None are from this chapter. However, you and I are both slaves to the system. So here are the topics that don't fit anywhere else.

Let's start at the beginning, counting.

COUNTING

The **basic law of counting** says: "If you can do something in p ways, and a second thing in q ways, and a third thing in r ways, and so on, the total number of ways you can do the first thing, then the second thing, then the third thing , etc., is $p \times q \times r \times \ldots$.

Example 1: If we have a lunch choice of 5 sandwiches, 4 desserts, and 3 drinks, and we can have one of each, how many different meals could we choose?

Solution: We can choose from $(5)(4)(3) = 60$ different meals

Arrangements

Let $n(A)$ be the number of elements in set A. In how many ways can these elements be arranged? The answer is that the first has n choices, the second has $(n - 1)$ choices (because one is already used), the third has $(n - 2)$ choices, all the way down to the last element, which has only one choice. In general, if there are n choices, the number of ways to choose is $n!$ (read as "**n factorial**") $= n(n - 1)(n - 2) \times \ldots (3)(2)(1)$.

Example 2: In how many ways can five people line up?

Solution: This is just $5 \times 4 \times 3 \times 2 \times 1 = 120$.

Note *Write $5 \times 4 \times 3 \times 2 \times 1$ as $5!$, read "5 factorial." In addition, $0! = 1$ since this definition makes all the formulas true! (Not factorial.)*

Example 3: How many ways can 5 people sit in a circle?

Solution: It would appear to be the same question as Example 2, but it's not. If we draw the picture, each of the five positions would be the same. The answer is $(5)(4)(3)(2)(1) \div 5 = (4)(3)(2)(1) = 24$. So n people can sit in a circle in $(n - 1)!$ ways.

Permutations

Permutations are essentially the law of counting without repeating, but order is important.

Example 4: How many ways can 7 people occupy 3 seats on a bench?

Solution: Any one of 7 people can be in the first seat, then any one of 6 people can be in the second seat, and any one of 5 people can be in the third seat. The total number would be $(7)(6)(5) = 210$ ways. There are many notations for permutations. One notation for this example would be $P(7, 3)$. Another symbol is $P_{7,3}$.

In general $P(n, r) = \dfrac{n!}{(n - r)!}$.

So $P(7, 3) = \dfrac{7!}{(7 - 3)!} = \dfrac{7 \times 6 \times 5 \times 4 \times 3 \times 2 \times 1}{4 \times 3 \times 2 \times 1} = 7 \times 6 \times 5.$

Combinations

Combinations are essentially the law of counting, with no repetition, and order doesn't matter.

Example 5: How many sets of three different letters can be made from eight different letters?

Solution: Because order doesn't matter, unlike the case for permutations, *AB* is the same as *BA*. So we can take the number of permutations, but we have to divide by the number of duplicates. It turns out that the duplicates for 3 letters is $3 \times 2 \times 1 = 6$. So we would have $\frac{8 \times 7 \times 6}{3 \times 2 \times 1} = 56$.

Again, there are many notations for combinations; the most common are $C(n, r)$, $C_{n,r}$, and $\binom{n}{r}$.

In general, $C(n, r) = \frac{n!}{r!(n-r)!}$.

So $C(8, 3) = \frac{8!}{3!5!} = \frac{8 \times 7 \times 6 \times 5 \times 4 \times 3 \times 2 \times 1}{3 \times 2 \times 1 \times 5 \times 4 \times 3 \times 2 \times 1} = \frac{8 \times 7 \times 6}{3 \times 2 \times 1} = 56$.

Usually, when the words "set" or "committee" are used, it implies combinations. If an element is in a set, order doesn't matter. If you are on a committee, it doesn't matter the order. You're stuck on the committee.

Avoiding Duplicates

When we count how many ways to do *A* or *B*, we should be careful not to count any item twice. We must subtract any items that include both *A* and *B*:

$$N(A \text{ or } B) = N(A) + N(B) - N(A \text{ and } B)$$

Example 6: Thirty students take French or German. If 20 took French and 18 took German, and if each student took at least one language, how many took both French and German?

Solution: $N(A \text{ or } B) = N(A) + N(B) - N(A \text{ and } B)$ or $30 = 20 + 18 - x$, so $x = 8$ took both languages.

Example 7: Forty students take Chinese or Japanese. If 9 take both and 20 take Japanese, how many students take Chinese?

Solution: $N(C \text{ or } J) = N(C) + N(J) - N(\text{both})$, or $40 = x + 20 - 9$, so $x = 29$ take Chinese.

Let's do some multiple-choice exercises.

For Exercises 1–5, use the set {*e, f, g, h, i*}. A word is considered to be any group of letters together; for example, hhg is a three-letter word.

Exercise 1: From this set, the number of three-letter words is:

A. 6
B. 27
C. 30
D. 60
E. 125

Exercise 2: How many three-letter permutations are there in this set?

A. 6
B. 27
C. 30
D. 60
E. 125

Exercise 3: How many three-letter words starting with a vowel and ending in a consonant can be made from this set?

A. 6
B. 27
C. 30
D. 60
E. 125

Exercise 4: How many three-letter words with the second and third letters the same can be made from this set?

A. 5
B. 20
C. 25
D. 60
E. 125

Exercise 5: How many three-letter permutations with the first and last letters *not* vowels can be made from this set?

A. 18
B. 27
C. 30
D. 45
E. 125

Exercise 6: Fifty students take Spanish or Portuguese. If 20 take both and 40 take Spanish, the number of students taking Portuguese *only* is

A. 0
B. 5
C. 10
D. 20
E. 30

Let's look at the answers.

Answer 1: **E:** $(5)(5)(5) = 125$.

Answer 2: **D:** $(5)(4)(3) = 60$.

Answer 3: **C:** The first letter has 2 choices, the second can be any 5, and the third has 3 choices, so $(2)(5)(3) = 30$.

Answer 4: **C:** There are 5 choices for the first two letters, but there is only 1 choice for the third letter because it must be the same as the second, so $(5)(5)(1) = 25$.

Answer 5: **A:** There are 3 choices for the first letter, but only 2 choices for the last letter because it can't be a vowel and must be different than the first letter. There are three choices for the middle letter because two letters have already been used, so the answer is $(3)(3)(2) = 18$. These questions must be read very carefully!

Answer 6: **C:** This is not quite the same as the previous exercise. $N(S \text{ or } P) = N(S) + N(P) - N(\text{both})$; $50 = 40 + x - 20$; $x = 30$. But that is not the answer. If 30 take Portuguese and 20 take both, then 10 take Portuguese only.

PROBABILITY

The probability of an event is the number of "good" outcomes divided by the total number of outcomes possible, or $Pr(\text{success}) = \dfrac{\text{good outcomes}}{\text{total outcomes}}$.

Example 8: Consider the following sets: {26-letter English alphabet}; vowels = $\{a, e, i, o, u\}$; consonants = {the rest of the letters}. What are the probabilities of choosing a vowel? a consonant? any letter? π?

Solutions: $Pr(\text{vowel}) = \dfrac{5}{26}$; $Pr(\text{consonant}) = \dfrac{21}{26}$; $Pr(\text{letter}) = \dfrac{26}{26} = 1$; $Pr(\pi) = \dfrac{0}{26} = 0$.

Probability follows the same rule about avoiding duplicates as discussed in the previous section.

$$Pr(A \text{ or } B) = Pr(A) + Pr(B) - Pr(A \text{ and } B)$$

Example 9: What is the probability that a spade or an ace is pulled from a 52-card deck?

Solution: $Pr(\text{Spade or ace}) = Pr(\text{Spade}) + Pr(\text{Ace}) - Pr(\text{Spade ace}) =$

$$\frac{13}{52} + \frac{4}{52} - \frac{1}{52} = \frac{16}{52} = \frac{4}{13}.$$

As weird as it sounds, whenever I taught this in a class, I never failed to have at least two students who didn't know what a deck of cards was, and I taught in New York City!

Use this figure for Examples 10 and 11. In the jar are five red balls and three yellow balls.

Example 10: What is the probability that two yellow balls are picked, with replacement?

Solution: $Pr(\text{2 yellow balls, with replacement}) = \left(\frac{3}{8}\right)\left(\frac{3}{8}\right) = \frac{9}{64}$

Example 11: What is the probability of picking two yellow balls, without replacement?

Solution: $Pr(\text{2 yellow balls, no replacement}) = \left(\frac{3}{8}\right)\left(\frac{2}{7}\right) = \frac{3}{28}$

SETS

We have mentioned the topic of sets informally. Now let's be a little more formal.

Sets are denoted by braces $\{a, b, c\}$. This set has three elements: a, b, and c.

Example 12: Let $A = \{a, b\}$, $B = \{b, a\}$, $C = \{a, b, b, b, a, b, b, a, a\}$. How is set A related to set B? To set C?

Solutions: $A = B$ because the order does not matter in sets. $A = C$ because repeated elements are counted only once; C has only two elements in it: a and b.

We write $A \cup B$, read "A **union** B," as the set of elements in A or in B or in both.

We write $A \cap B$, read "A **intersect(ion)** B," as the set of elements common to A and B.

The **null set**, written { } or ϕ (the Greek letter phi), is the set with nothing in it.

Example 13: Let $D = \{a, b, c, d, e, f\}$, $E = \{c, d, f, g\}$, $F = \{a, b, e\}$. Find $D \cup E$, $D \cap E$, and $E \cap F$.

Solutions: $D \cup E = \{a, b, c, d, e, f, g\}$; $D \cap E = \{c, d, f\}$; $E \cap F = \phi$. Sets with no common elements are called **disjoint**.

We say A is a subset of B, written $A \subseteq B$ if every element in A is also in B. If a set has n elements, it has 2^n subsets.

Example 14: Write all the subsets of $G = \{a, b, c\}$.

Solution: Because G has 3 elements, there are $2^3 = 8$ subsets. They are: ϕ, $\{a\}$, $\{b\}$, $\{c\}$, $\{a, b\}$, $\{a, c\}$, $\{b, c\}$, $\{a, b, c\}$.

Two other terms relating to sets that we need to know are:

- The **universe**, which is the set of everything we are talking about.

The universe could be U = {all animals}, or it could be A = {all mammals.} It depends on what our topic of interest is. Note that $A \subseteq U$.

- A **complement** is the set of elements in the universe not in a specific set. The complement is denoted by a superscript letter c, so A^c is the complement of A. There are many other symbols in math for the complement.

For example, if U = {all animals} and M = {all mammals}, then M^c = {reptiles, insects, fish, etc.}

Example 15: If the alphabet is the universe, and A = {vowels}, what is A^c?

Solution: A^c = {consonants}.

Note *If we change the universe, we also automatically change the complement. This type of complement is spelled with two e's; not as "compliment," which is flattering.*

The next topic is not on the ACCUPLACER, but it is an introduction to determinants, a topic that *is* on the ACCUPLACER.

MATRICES

A matrix is an array of numbers. We will use letters to stand for numbers. $\begin{bmatrix} a & b & c \\ d & e & f \end{bmatrix}$ is a 2 by 3 matrix, written 2×3. It has 2 rows: row 1, $\begin{bmatrix} a & b & c \end{bmatrix}$; and row 2, $\begin{bmatrix} d & e & f \end{bmatrix}$. It has 3 columns: column 1, $\begin{bmatrix} a \\ d \end{bmatrix}$; column 2, $\begin{bmatrix} b \\ e \end{bmatrix}$; and column 3, $\begin{bmatrix} c \\ f \end{bmatrix}$.

To add two matrices, they must have the same number of rows and columns. You then add the corresponding entries.

Example 16: $\begin{bmatrix} a & b & c \\ d & e & f \end{bmatrix} + \begin{bmatrix} r & s & t \\ u & v & w \end{bmatrix} =$

Solution: $\begin{bmatrix} a+r & b+s & c+t \\ d+u & e+v & f+w \end{bmatrix}$

To multiply two matrices, the number of columns of the first matrix must be the same as the number of rows of the second. If we multiplied a 2×5 matrix by a 5×3 matrix, we would get a 2×3 matrix. However, multiplying a 5×3 matrix by a 2×5 matrix is impossible because 3 (columns in the first matrix) is not equal to 2 (rows in the second matrix).

The next example shows the procedure for multiplying matrices.

Example 17: Let $A = \begin{bmatrix} b & c & d \\ f & g & h \end{bmatrix}$ and $B = \begin{bmatrix} j & m \\ k & n \\ l & p \end{bmatrix}$. Find AB and BA.

Solutions: A is a 2×3 matrix, and B is a 3×2 matrix. AB would be a 2×2 matrix.

The first entry is found by multiplying the first row of A by the first column of B. The entry would be the element in the first row, first column of AB. We get $bj + ck + dl$. The first row of A multiplied by the second column of B gives the first row, second column of AB. We get $bm + cn + dp$. The second row, first column of AB is $fj + gk + hl$, and the second column, second row of AB is $fm + gn + hp$. The matrix AB would look like this:

$$AB = \begin{bmatrix} bj + ck + dl & bm + cn + dp \\ fj + gk + hl & fm + gn + hp \end{bmatrix}$$

Similarly, to get BA, we are multiplying a 3×2 matrix by a 2×3 matrix, so we get a 3×3 matrix:

$$BA = \begin{bmatrix} jb + mf & jc + mg & jd + mh \\ kb + nf & kc + ng & kd + nh \\ lb + pf & lc + pg & ld + ph \end{bmatrix}$$

This next topic is required by the ACCUPLACER.

DETERMINANTS

If we have a square matrix (the number of rows equals the number columns), we can give the matrix a value.

A 2–by–2 determinant, written 2×2, looks like this: $\begin{vmatrix} a & b \\ c & d \end{vmatrix}$. Its value is $(ad - bc)$.

Example 18: $\begin{vmatrix} 4 & 5 \\ 6 & -7 \end{vmatrix} =$

Solution: $4(-7) - 5(6) = -58.$

Note *The value of a determinant can be positive, negative, or zero.*

Determinants can be used to solve two equations in two unknowns by the following method.

Suppose we have $ax + by = m$

$cx + dy = n$

To solve for x and y, we would use the following determinants:

$$x = \frac{\begin{vmatrix} m & b \\ n & d \end{vmatrix}}{\begin{vmatrix} a & b \\ c & d \end{vmatrix}} \text{ and } y = \frac{\begin{vmatrix} a & m \\ c & n \end{vmatrix}}{\begin{vmatrix} a & b \\ c & d \end{vmatrix}}.$$

The denominators are the coefficients of the x and y terms. If you are solving for x, replace the coefficients of the x numbers in the numerator by m and n from the answer column. Likewise, when you solve for y, the m and n replace the coefficients of the y numbers.

Example 19: Solve for x and y by using determinants:

$5x - 7y = 29$

$11x + 9y = 10$

Solution:

$$x = \frac{\begin{vmatrix} 29 & -7 \\ 10 & 9 \end{vmatrix}}{\begin{vmatrix} 5 & -7 \\ 11 & 9 \end{vmatrix}} = \frac{(29)(9) - (-7)(10)}{(5)(9) - (-7)(11)} = \frac{331}{122}$$

$$y = \frac{\begin{vmatrix} 5 & 29 \\ 11 & 10 \end{vmatrix}}{\begin{vmatrix} 5 & -7 \\ 11 & 9 \end{vmatrix}} = \frac{(5)(10) - (29)(11)}{(5)(9) - (-7)(11)} = -\frac{269}{122}.$$

Note *The beauty of this method is that if the numbers are really messy, there is no problem finding the solution. Any other way, the arithmetic would be much worse.*

Note *If the ACCUPLACER asks you to solve two equations in two unknowns, you may do it any way you want since the test is multiple choice.*

You can similarly use determinants for three equations in three unknowns. The following method shows how to evaluate a 3×3 determinant. The ACCUPLACER does not ask to solve three equations in three unknowns.

To evaluate $\begin{vmatrix} a & b & c \\ d & e & f \\ g & h & i \end{vmatrix}$, rewrite the first two columns as $\left|\begin{array}{ccc|cc} a & b & c & a & b \\ d & e & f & d & e \\ g & h & i & g & h \end{array}\right|$.

The value is the product of the diagonals going from the upper left to the lower right minus the product of the diagonals going from the lower left to the upper right.

In symbols, the value is $aei + bfg + cdh - gec - hfa - idb$.

Example 20: Evaluate $\begin{vmatrix} 4 & 5 & 0 \\ 2 & -1 & 1 \\ 0 & 3 & -3 \end{vmatrix}$.

Solution: $\left|\begin{array}{ccc|cc} 4 & 5 & 0 & 4 & 5 \\ 2 & -1 & 1 & 2 & -1 \\ 0 & 3 & -3 & 0 & 3 \end{array}\right| =$

$$(4)(-1)(-3) + 5(1)(0) + 0(2)(3) - 0(-1)(0) - (3)(1)(4) - (-3)(2)(5) = 30.$$

Facts about determinants:

- If one row (or column) is all zeroes, the value of the determinant is 0.
- If two rows (or columns) are equal, the value of the determinant is 0.
- If two rows (or columns) are interchanged, the value is multiplied by -1.

Note *There are two methods of evaluating a 3×3 determinant. This method is the easier, but it will not work for 4×4, 5×5, or higher-order determinants.*

TRANSLATIONS

Let's look at the graph of $y = x^2$:

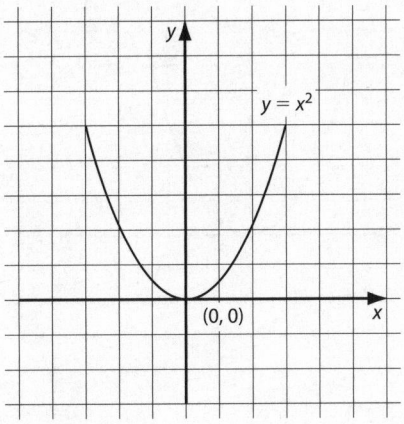

Translations of this graph include the following:

- The same graph 3 units down would be $y = x^2 - 3$.

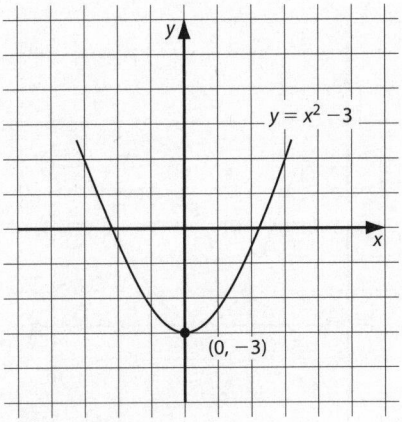

- Or the same graph 5 units up would be $y = x^2 + 5$.

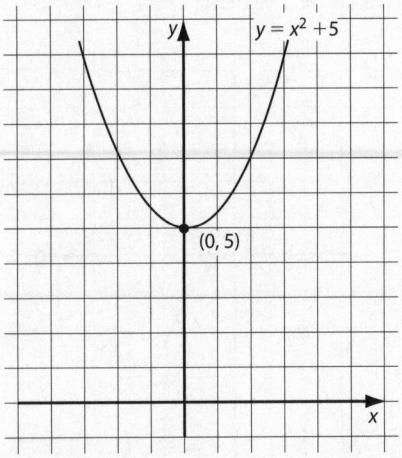

- Or the same graph 4 units to the left would be $y = (x + 4)^2$.

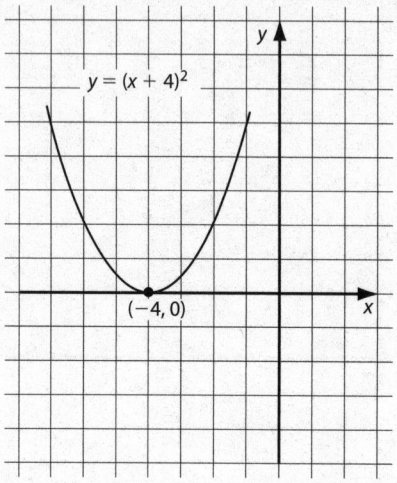

- Or the same graph 9 units to the right would be $y = (x - 9)^2$.

Note *Each box represents 2 units.*

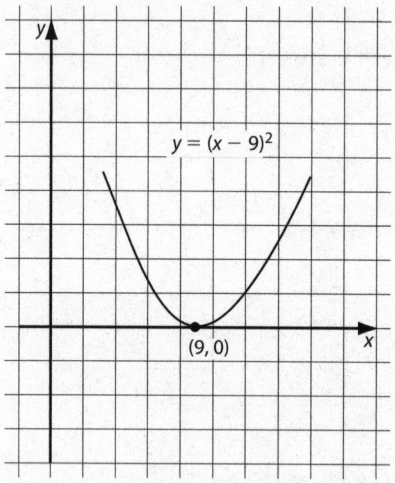

- Finally, the same graph upside down would be $y = -x^2$.

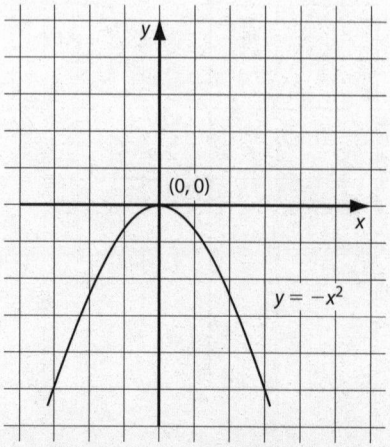

Example 21: What would the curve look like if it were 4 units to the left, 3 units down, and upside down?

Solution: The equation would be $y = -(x + 4)^2 - 3$.

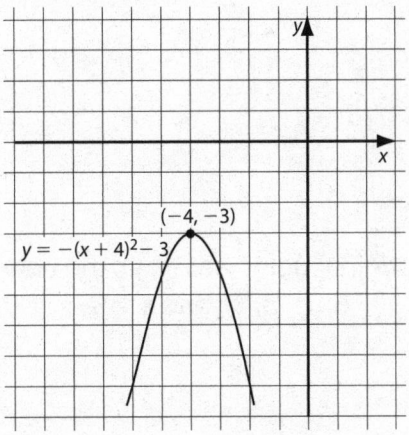

Note *The higher the positive coefficient in front the x^2 term, the quicker the curve goes up. (Conversely, the higher the absolute value of the negative coefficient in front of the x^2 term, the quicker the curve goes down.)*

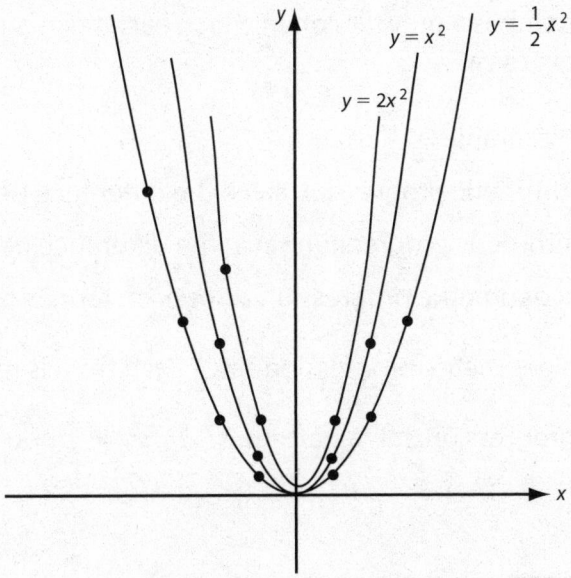

Note *Notice that with trig functions, we did the same thing.*

Q Let's try a multiple-choice exercise with translations.

Exercise 7: After moving 4 units to the left and down 6 units, a point on a graph ended up at (3, 3). The original point was

A. (7, 9)
B. (−1, 9)
C. (−1, −3)
D. (7, −3)
E. none of these

A Let's look at the answer.

Answer 7: **A:** If the original point was (x, y), the new point would be $(x - 4, y - 6)$. So $(x - 4, y - 6) = (3, 3)$, and $x = 7, y = 9$.

PROGRESSIONS

There are two kinds of progressions:

1. An **arithmetic progression** is a progression for which the difference between any two neighboring terms is the same.
2. A **geometric progression** is a progression in which each term is multiplied by the same number to get the next term.

Let's look at the following examples:

- 9, 13, 17, 21… is an arithmetic progression since the difference between terms is 4.
- 4, 3.5, 3, 2.5… is an arithmetic progression since the difference between terms is −0.5.
- 5, −10, 20, −40… is a geometric progression since each term is multiplied by −2.
- 486, 162, 54, 18… is a geometric progression since each term is multiplied by $\frac{1}{3}$.
- 1, 4, 9, 16…. is not a progression. It is a pattern: $1^2, 2^2, 3^2, 4^2, \ldots$ but does not fit either of our two patterns.

Arithmetic Progressions

Let d = the difference between terms. The first term is called either a or a_1. The second term is $a_2 = a + d$. The third term is $a + 2d$; the fourth term is $a + 3d$; and the last (or nth) term, denoted by ℓ or a_n is $a + (n - 1)d$.

The derivation of the sum is adorable, so let's do it.

If we start from the first number, the sum $S = a + (a + d) + (a + 2d) + (a + 3d) + \ldots + [a + (n - 1)d]$.

If we start from the last number, the sum $S = \ell + (\ell - d) + (\ell - 2d) + (\ell - 3d) + \ldots + [\ell - (n - 1)d]$.

Adding these two equations, we see that all the d terms drop out, and we are left with n terms, $2S = n(a + \ell)$.

So $S = \left(\frac{n}{2}\right)(a + \ell)$. Substituting $\ell = a + (n - 1)d$, we get $S = \left(\frac{n}{2}\right)(2a + (n - 1)d)$.

Summary: There are three formulas you need to know:

$\ell = a + (n - 1)d$; $S = \left(\frac{n}{2}\right)(a + \ell)$; and $S = \left(\frac{n}{2}\right)(2a + (n - 1)d)$.

Example 22: Find the 20th term if the first term is 5 and the difference is 3.

Solution: $\ell = a + (n - 1)d$; so $\ell = 5 + (20 - 1)3 = 62$.

Example 23: Find the sum if the first term is 6, the last term is 7, and the number of terms is 10.

Solution: $S = \left(\frac{n}{2}\right)(a + \ell)$; $S = \left(\frac{10}{2}\right)(6 + 7) = 65$.

Example 24: Find the sum if the first term is -3, the difference is -7, and the number of terms is 8.

Solution: $S = \left(\frac{n}{2}\right)(2a + (n - 1)d)$; $S = \left(\frac{8}{2}\right)[2(-3)+(8-1)(-7)] = -220$.

Example 25: An arithmetic sequence is 7, ____, ____, ____, 55. Fill in the missing terms.

Solution: We have a 5-term arithmetic series; $a = 7$, and the last term $a_5 = a + 4d = 55$; so $7 + 4d = 55$; $4d = 48$; and $d = 12$. The terms are 7, 7 + 12, 7 + 24, 7 + 36, and 7 + 48, or 7, 19, 31, 43, and 55.

Example 26: If the sum is 10, the first term is -2, and the difference is 3, find the number of terms.

Solution: A toughie! We use the formula $S = \left(\frac{n}{2}\right)(2a + (n - 1)d)$; so

$10 = \left(\frac{n}{2}\right)[2(-2) + (n - 1)(3)]$. Multiply through by 2, combine terms to get $20 = n(3n - 7)$, or $3n^2 - 7n - 20 = (3n + 5)(n - 4) = 0$. The number of terms must be a positive integer, so $n = 4$, and there are 4 terms.

Geometric Progressions (Finite)

A geometric progression has a common ratio, r. Again, we let a = the first term. The second term is then ar; the third term is ar^2; the fourth term is ar^3; and the last term is $\ell = a_n = ar^{n-1}$.

Again, the proof of the sum is adorable. So let's do it.

We write out the sum. $S = a + ar + ar^2 + ar^3 + \ldots + ar^{n-1}$.

If we multiply by $-r$, we get $-rS = -ar - ar^2 - ar^3 - \ldots. - ar^{n-1} - ar^n$.

If we now add these two equations, all of the middle terms add to 0, we get

$S - rS = S(1 - r) = a - ar^n = a(1 - r^n)$.

So $S = \dfrac{a - ar^n}{1 - r}$ or $\dfrac{a(1 - r^n)}{1 - r}$. Substituting $\ell = ar^{n-1}$ in the first expression, we get $S = \dfrac{a - r\ell}{1 - r}$.

In summary,

$\ell = ar^{n-1}$; $S = \dfrac{a(1 - r^n)}{1 - r}$ or $S = \dfrac{a - ar^n}{1 - r}$ or $S = \dfrac{a - r\ell}{1 - r}$.

Note *The first form for S is the one that is usually used.*

Example 27: Suppose the first term of a geometric progression is 10 and the ratio is 3; find the fifth term.

Solution: We have $\ell = ar^{n-1}$; so $\ell = (10)(3)^4 = 810$.

Example 28: Find the sum for the geometric progression of Example 27.

Solution: $S = \dfrac{a(1 - r^n)}{1 - r} = \dfrac{10(1 - 3^5)}{1 - 3} = 1210$.

Example 29: We have a 3-term geometric sequence 6, ____, 11. Find the missing term.

Solution: We can set up a proportion here: $\dfrac{6}{x} = \dfrac{x}{11}$; $x^2 = 66$; $x = \pm\sqrt{66}$. These two numbers are sometimes called the **geometric mean**.

Infinite Geometric Series

Suppose the geometric series goes on forever. Let's see what would happen to the sum. Let's look at the possibilities:

- If $r > 1$: Look at the series 4, 40, 400, 4000,.... Clearly, if you added all this up, the sum would be infinite.
- If $r < -1$: Look at the series 10, -20, 40, -80,.... If you stopped after a negative term the sum would be very small (a large negative number). If you stopped after a positive number, the sum would be big. Clearly, there is no one sum.
- If $r = 1$: Look at the series 7, 7, 7, 7,.... The sum of an infinite number of 7's is infinite.
- If $r = -1$: Look at the series 8, -8, 8, -8, 8, -8,.... If you stopped after an odd number, the sum is 8; if you stopped after an even number of terms, the sum is 0. The sum must tend toward one number as n increases, so $r = -1$ is not allowed.
- If $-1 < r < 1$: Let's take $r = \frac{1}{10}$, for example. Let's look at $\left(\frac{1}{10}\right)^n$ as n gets very big. The sequence becomes $\frac{1}{10}, \frac{1}{100}, \frac{1}{1{,}000}, \frac{1}{10{,}000}, \ldots$. The bigger n gets, the smaller the numbers get. The limit is $r^n = 0$. So the finite geometric sum, $S = \frac{a(1 - r^n)}{1 - r}$, becomes the infinite geometric sum $S = \frac{a}{1 - r}$.

Example 30: Find the infinite sum of 100, 80, 64,....

Solution: We see $a = 100$ and $r = \frac{80}{100} = \frac{4}{5}$. $S = \frac{a}{1 - r} = \frac{100}{1 - \frac{4}{5}} = 500$.

Example 31: Change the repeating decimals .3333... and .979797.... to rational numbers.

Solution: We can write .3333... as $.3 + .03 + .003 + \ldots$; $a = .3$ and $r = .1$; so $S = \frac{a}{1 - r} = \frac{.3}{1 - .1} = \frac{.3}{.9} = \frac{1}{3}$. Similarly, we can write .979797... as $.97 + .0097 + .000097 + \ldots$; so $a = .97$ and $r = .01$; so $S = \frac{.97}{1 - .01} = \frac{97}{99}$.

CONIC SECTIONS

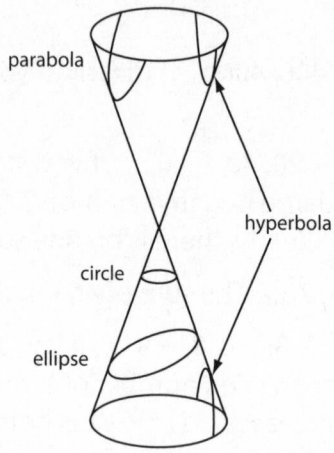

As the figure above shows, the conic sections are the intersection of a more general circular cone with a plane. If it is parallel to the top or bottom, the figure is a circle. If the plane cuts at an angle to the base, we have an ellipse. If the plane cuts parallel to the line pictured, we get a parabola. If we have a plane cutting the top and bottom perpendicular to both, we get a hyperbola.

In this section, we talk a little more about the circle. We then talk about a second way of looking at the parabola. However, as illogical as it is, I (and several other authors as well) have found that if we talk about ellipses and hyperbolas before talking about parabolas, students learn better than if parabolas are talked about first. But first let's talk a little more about circles.

Circles (Continued)

We will pick up from Chapter 9, in which we learned that the general form of a circle with center at (h, k) and radius r is

$(x - h)^2 + (y - k)^2 = r^2$.

So $x^2 + y^2 = 100$ would be a circle with its center at the origin and radius 10. Now suppose we have the circle $(x - 6)^2 + (y + 2)^2 = 100$. Has the size of the circle changed? No! The radius is still 10; however, the circle as been moved (translated) 6 units to the right and 2 units down.

What if we were given the circle in multiplied-out form? In order to put it in general form, we have to review how to "complete the square."

Recall that $(x + k)^2 = x^2 + 2kx + k^2$. Let's translate this into English. If the coefficient of x^2 is 1, to complete the square, take half the coefficient of x and square it and add it to the expression to make a perfect square. So to complete the square for $x^2 - 10x$, we would add 25:

$$x^2 - 10x + \left(\frac{-10}{2}\right)^2 = x^2 - 10x + 25 = (x - 5)(x - 5) = (x - 5)^2.$$

Now, let's see the usefulness of completing the square in the next example.

Example 32: Find the radius and center of the circle $2x^2 + 2y^2 + 12x - 16y + 20 = 0$.

Solution: First, we must divide by 2 to make the coefficients of the squared terms 1, and we get $x^2 + 6x + y^2 - 8y + 10 = 0$. We then group like letters together. Now we complete the squares for the x and y terms separately, remembering that whatever is added to one side of an equation must also be added to the other side. We then have

$$x^2 + 6x + \left(\frac{6}{2}\right)^2 + y^2 - 8y + \left(\frac{-8}{2}\right)^2 = -10 + 9 + 16.$$

Simplifying, we have $(x + 3)^2 + (y - 4)^2 = 15$. This circle has center $(-3, 4)$ and radius $\sqrt{15}$.

Note *The way we identify an equation as being a circle is that the coefficients of x^2 and y^2 are the same number and the same sign. After completing the square, the number on the right side must be positive. If it is 0, it is a point. If it is a negative number, then the answer is the null set, no graph.*

Ellipses

An ellipse has as its basis two points, called foci (singular: *focus*), with coordinates $(c, 0)$ and $(-c, 0)$, which do not appear on the curve of the ellipse itself. Rather, the sum of the distances from one focus to any point (x, y) on the curve and from that point to the other focus is a constant; let's call that sum $2a$. This curve forms the ellipse.

If you want to draw an ellipse, put two thumbtacks at points $(c, 0)$ and $(-c, 0)$. Then find a non-elastic string of length greater than $2c$, and form a loop with the string. Loop the string over the tacks and place a pencil in that loop and stretch it out to the max. Pull the pencil around the tacks 360°. This will sketch an **ellipse**.

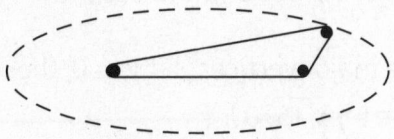

We will not derive the equation for an ellipse. The equation looks easy, but the algebra getting it isn't.

The general equation for an ellipse is

$$\frac{x^2}{a^2} + \frac{y^2}{b^2} = 1, \text{ where } a^2 - b^2 = c^2.$$

The coefficients of x^2 and y^2 are the same sign but different numbers.

To understand the ellipse, we must define its **center**, which is halfway between the two foci. In addition, the line through the foci and the center to the ellipse is the **major axis**, and a line perpendicular to the major axis through the center to the ellipse is the **minor axis**.

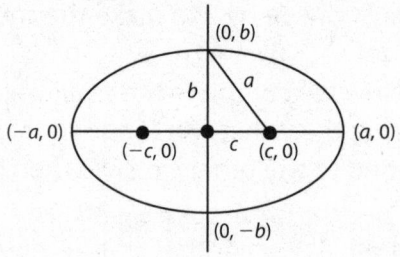

Now let's identify all the letters, all vertices, and the foci as shown in the figure:

- a is the length of the semi-major axis, or half the major axis.
- b is the length of the semi-minor axis, or half the minor axis.
- c is half the distance between the foci.
- $(\pm a, 0)$ are the major vertices.
- $(0, \pm b)$ are the minor vertices.
- $(\pm c, 0)$ are the foci.

Note *a is also the distance from a focus to a minor vertex.*

In addition, the shape of an ellipse is described by its eccentricity. The **eccentricity** $e = \dfrac{c}{\text{length of the semi-major axis}}$. In the case of a circle, $e = 0$. As e gets closer to 1, the ellipse gets longer and less wide. When $e = 1$, the curve is a parabola. When $e > 1$, as we shall see, the curve is a hyperbola. Parabolas and hyperbolas are discussed later in this chapter.

Example 33: Find all vertices, foci, and the eccentricity for $\dfrac{x^2}{15} + \dfrac{y^2}{8} = 1$.

Solution: To find the major vertices, set $y = 0$; then $x^2 = 15$; $x = \pm\sqrt{15}$. The major vertices are $(\pm\sqrt{15}, 0)$.

To find minor vertices, set $x = 0$; then $y^2 = 8$; $y = \pm\sqrt{8}$. The minor vertices are $(0, \pm\sqrt{8})$.

$a^2 - b^2 = c^2$; $15 - 8 = 7$. So the foci are $\left(\pm\sqrt{7}, 0\right)$.

$$e = \frac{c}{\text{length of the semi-major axis}} = \frac{\sqrt{7}}{\sqrt{15}}.$$

Example 34: Find all vertices, foci, and eccentricity for $11x^2 + 3y^2 = 66$.

Solution: First, we know this is an ellipse because the coefficients of the x^2 and y^2 terms are the same sign but different numbers. We first divide by 66.

We get $\frac{x^2}{6} + \frac{y^2}{22} = 1$. Since the number under the y^2 term is larger, the major axis is longer in the y direction. (Only in the case of an ellipse is this true.)

Letting $x = 0$, the major vertices are $\left(0, \pm\sqrt{22}\right)$.

Letting $y = 0$, the minor vertices are $\left(\pm\sqrt{6}, 0\right)$.

Then $c = \sqrt{22 - 6} = 4 : (0, \pm 4)$.

Finally, $e = \frac{4}{\sqrt{22}}$.

Example 35: Find all the appropriate information for the ellipse

$$\frac{(x-5)^2}{6} + \frac{(y+3)^2}{22} = 1.$$

Solution: This is the same as example 34, except the center of the ellipse is no longer at (0, 0). But similar to the case for the circle, we see the new center is at (5, −3). So all of the points are translations from (5, −3).

The major vertices are $\left(5, -3 \pm \sqrt{22}\right)$.

The minor vertices are $\left(5 \pm \sqrt{6}, -3\right)$.

The foci are $(5, -3 \pm 4)$.

The eccentricity is, of course, the same since the shape is the same.

Example 36: Find the center, all vertices, foci, and eccentricity of $5x^2 + 7y^2 - 80x + 56y + 397 = 0$.

Solution: We cannot complete the square as we did for the circle. If we divide by 5, the coefficient of the y^2 term becomes $\frac{7}{5}$. Similarly, we can't divide by 7. So for the x terms we will factor out a 5 (and for the y terms we will factor out a 7) and complete the squares in the parentheses.

We have $5(x^2 - 16x + 64) + 7(y^2 + 8y + 16) = 5(64) + 7(16) - 397$.

So we have $5(x - 8)^2 + 7(y + 4)^2 = 35$. Dividing by 35, we get

$$\frac{(x-8)^2}{7} + \frac{(y+4)^2}{5} = 1.$$

So the center is at $(8, -4)$; the major vertices are $\left(8 \pm \sqrt{7}, -4\right)$; the minor vertices are $\left(8, -4 \pm \sqrt{5}\right)$; the foci are $\left(8 \pm \sqrt{2}, -4\right)$, and the eccentricity is $e = \frac{\sqrt{2}}{\sqrt{7}} = \frac{\sqrt{14}}{7}$.

Note *The coefficient of x^2 is 7 and the coefficient of y^2 is 5. After completing the square, the number on the right will be 35 or a multiple of 35 (5×7). It doesn't have to be, but if it isn't, the arithmetic is really, really bad.*

The next conic section we'll do is the hyperbola; again, we will look at four examples, just like we did for the ellipse. However, hyperbolas have asymptotes. I feel it is necessary to do a little explaining about them first.

Asymptotes

For our purposes, the following is the definition of an asymptote.

Asymptote: A line to which a curve gets "very close at the end" but never hits.

Lots of notes:

Note *"At the end" means when the curve goes to $\pm\infty$.*

Note *"Very close" is very complicated. We'll assume we all understand what this means.*

Note *An asymptote is a straight line approached by a curve that never intersects it.*

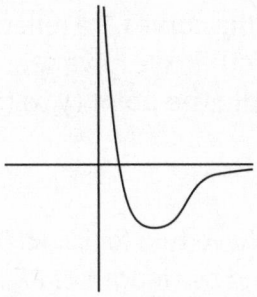

Notice that in the "middle" of the curve, the x-axis does not approximate the curve. But at the end, the curve gets very close to the x-axis but doesn't hit it. This is also true for the curve as it gets close to the y-axis but doesn't cross it. The axes (plural of axis) are the asymptotes in this figure.

Example 37: On the same set of axes, draw $y = 2^x$ and $y = \log_2 x$.

Solution:

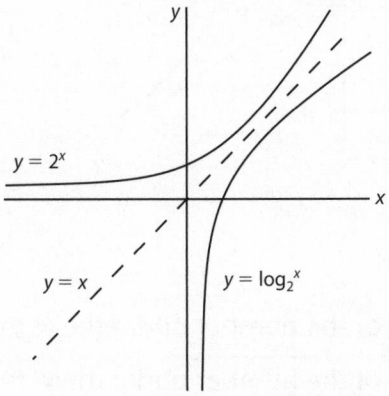

For $y = 2^x$, select the points $x = -3, -2, -1, 0, 1, 2, 3$. We get the points $\left(-3, \frac{1}{8}\right), \left(-2, \frac{1}{4}\right), \left(-1, \frac{1}{2}\right)$, (0, 1), (1, 2), (2, 4), and (3, 8).

Note *The x-axis is the asymptote for $y = 2^x$.*

Note *If we had to graph $y = 2^{x-10}$, the first number we take is $x = 10$, and then three values less than that and three values more.*

For $y = \log_2 x$, we use the definition $2^y = x$, and assign y values $-3, -2, -1, 0, 1, 2$, and 3.

We get the points $\left(\frac{1}{8}, -3\right), \left(\frac{1}{4}, -2\right), \left(\frac{1}{2}, -1\right)$, (1, 0), (2, 1), (4, 2), and (8, 3).

Note *The y-axis is the asymptote here.*

If we draw the line $y = x$, we see that the curves are reflections of each other. This graphically shows that they are **inverses** of each other. We have seen that the inverse of a function is made by changing every point (x, y) into the point (y, x) (changing the x and y values around).

Hyperbolas 1

In order to form a hyperbola, we first have two foci, $F_1(c, 0)$ and $F_2(-c, 0)$. Take a point $P(x, y)$ on the curve. Find the length of PF_1 and the length of PF_2. If $|PF_1 - PF_2|$, we get the equation $\frac{x^2}{a^2} - \frac{y^2}{b^2} = 1$, where $a^2 + b^2 = c^2$.

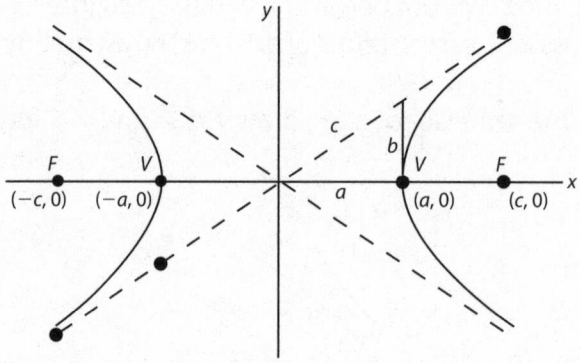

The points $(\pm a, 0)$ are called the transverse vertices.

The points $(\pm c, 0)$ are the foci.

The asymptotes are $y = \pm \dfrac{\text{the square root of the number under the } y^2 \text{ term}}{\text{the square root of the number under the } x^2 \text{ term}} x$.

The eccentricity is $e = \frac{c}{a}$.

Note *There is no* b *in the formulas for the vertices, foci, asymptotes, or eccentricity. If we draw a line tangent to the hyperbola at* ±a, *the length from that point of tangency to the asymptote is* b. *The distance from the origin to* (a, 0) *is* a. *So the "hypotenuse" of this unreal triangle is also* c, *equal to the distance from either focus to the origin.*

Example 38: Locate all information for the hyperbola $\frac{x^2}{3} - \frac{y^2}{7} = 1$.

Solution: We know this is a hyperbola because x^2 and y^2 have opposite signs. If $y = 0, x = \pm\sqrt{3}$, the transverse vertices are $(\pm\sqrt{3}, 0)$.

Since $a^2 + b^2 = c^2, 3 + 7 = 10 = c^2$. The foci are $(\pm\sqrt{10}, 0)$.

The asymptotes are $y = \pm\frac{\sqrt{7}}{\sqrt{3}}x = \pm\frac{\sqrt{21}}{3}x$.

The eccentricity is $e = \frac{c}{a} = \frac{\sqrt{10}}{\sqrt{3}} = \frac{\sqrt{30}}{3}$.

Note *If $x = 0$, y would be imaginary. That is why this is a left–right hyperbola.*

Note *The size of the numbers under the squared variables does* not *determine the shape; rather, it is the location of the minus sign.*

Example 39: $6x^2 - 7y^2 = -42$; give all the important information about this curve.

Solution: We know it's a hyperbola because the signs of the squared terms are different.

Dividing by -42, we get $\frac{y^2}{6} - \frac{x^2}{7} = 1$. If we drew this picture, it would be an up–down hyperbola because the x^2 term is the negative term.

The vertices are $(0, \pm\sqrt{6})$.

The foci are $(0, \pm\sqrt{13})$.

The asymptotes are $y = \pm\frac{\sqrt{6}}{\sqrt{7}}x = \pm\frac{\sqrt{42}}{7}x$.

The eccentricity is $e = \frac{\sqrt{13}}{\sqrt{6}} = \frac{\sqrt{78}}{3}$.

Example 40: Find the information for the graph of $\frac{(y-3)^2}{6} - \frac{(x+2)^2}{7} = 1$.

Solution: Just as in the case of the ellipse, all points are translations of the center (the point where the asymptotes cross). In this case, the center is at $(-2, 3)$. Remember, the x-coordinate always comes first. Due to the translation, the "x" in the asymptotes becomes $x + 2$, and the y becomes $y - 3$. This is an up-down hyperbola.

The vertices are $(-2, 3 \pm \sqrt{6})$. The foci are $(-2, 3 \pm \sqrt{13})$. The asymptotes are $y - 3 = \pm\frac{\sqrt{6}}{\sqrt{7}}(x + 2)$, or $y = 3 \pm \frac{\sqrt{42}}{7}(x + 2)$. The eccentricity is $e = \frac{\sqrt{13}}{\sqrt{6}} = \frac{\sqrt{78}}{6}$.

The last example here involves completing the square.

Example 41: Find all appropriate information for $4x^2 - 3y^2 - 16x + 18y - 23 = 0$.

Solution: We proceed as we did for the ellipse, except when you factor out $(x - 3)$ from the y terms, you must factor the minus sign out from both of them!

$4x^2 - 16x - 3y^2 + 18y = 23.$

$4(x^2 - 4x + 4) - 3(y^2 - 6y + 9) = 23 + 4(4) - 3(9).$

$4(x - 2)^2 - 3(y - 3)^2 = 12.$ We thus get $\dfrac{(x-2)^2}{3} - \dfrac{(y-3)^2}{4} = 1.$

The center is (2, 3). The vertices are $\left(2 \pm \sqrt{3}, 3\right)$. The foci are $\left(2 \pm \sqrt{7}, 3\right)$.

The eccentricity is $\dfrac{\sqrt{7}}{\sqrt{3}} = \dfrac{\sqrt{21}}{3}$. The asymptotes are $(y - 3) = \pm \dfrac{\sqrt{4}}{\sqrt{3}}(x - 2)$ or $y = 3 \pm \dfrac{2\sqrt{3}}{3}(x - 2)$.

Hyperbolas 2

There is another hyperbola that almost nobody mentions. However, you should know what it looks like.

It is of the form $xy = c$, where c is a constant. The asymptotes are the x- and y-axis.

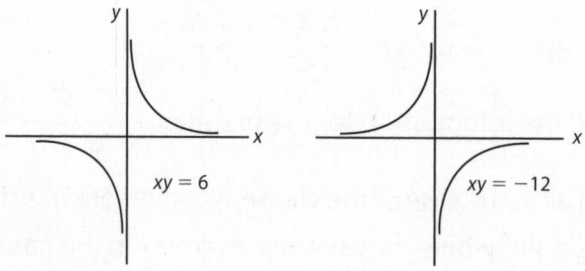

If $c > 0$, the graph is the left picture shown, such as $xy = 6$. If $c < 0$, the graph is the second one shown, such as $xy = -12$.

Note *You can graph these hyperbolas by taking* $x = \pm 1, \pm 2, \pm 3, \pm 6, \pm 12$.

Parabolas 2

We finally get to the second way of looking at parabolas, in addition to what we did in Chapter 14. The definition of parabola is:

A **parabola** is the set of all points equidistant (having the same distance) from a point (focus) and a line (directrix).

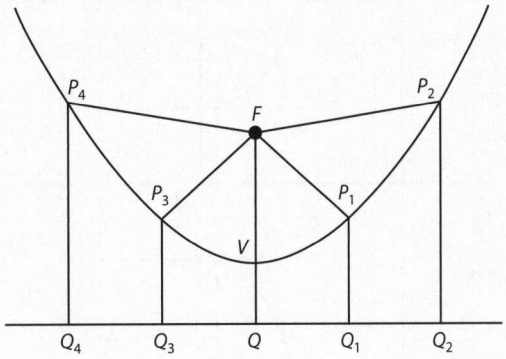

Let F = the focus and V = the vertex. We see that $FV = VQ$, $FP_1 = P_1Q_1$, $FP_2 = P_2Q_2$, etc. We see the parabola forming.

Note *Distance to a line (in this case, the directrix) always means perpendicular distance.*

The parabola has a very interesting property—it is the principle behind the parabolic microphone, radar, and the headlights of a car. My cousin Bobby married his bride Norma in a room that had a parabolic ceiling with the marriage ceremony at the focus. But the audience had to be careful. People standing at the focus hear everyone in the room. Sound waves go in both directions!

The equation for the parabola is short enough to derive.

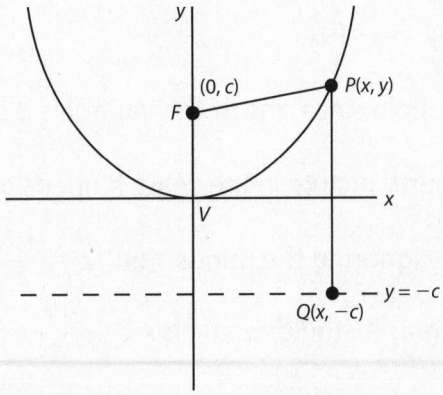

Let the focus be at the point $F(0, c)$, and the directrix be $y = -c$. Let $P(x, y)$ be on the parabola. Then $FP = PQ$. Q is on the directrix; so the y coordinate is $-c$. Q has the same x value as the point $P(x, y)$, so the coordinates of Q are $(x, -c)$. Saying $FP = PQ$ means $\sqrt{(x-0)^2 + (y-c)^2} = |y - (-c)|$.

Squaring both sides, we get $x^2 + y^2 - 2cy + c^2 = y^2 + 2cy + c^2$. Simplifying, we get $x^2 = 4cy$.

The following chart will help you relate the picture to the equation.

Vertex	Focus	Directrix	Shape	Equation
(0, 0)	(0, c)	$y = -c$		$x^2 = 4cy$
(0, 0)	(0, $-c$)	$y = c$		$x^2 = -4cy$
(0, 0)	(c, 0)	$x = -c$		$y^2 = 4cx$
(0, 0)	($-c$, 0)	$x = c$		$y^2 = -4cx$

Note *In some books the letter c is replaced by the letter p.*

Let's look at some examples for parabolas.

Example 42: Describe $y^2 = -19x$.

Solution: It is a parabola since one of the variables is squared and the other isn't.

It is the fourth picture in the chart; it opens to the left. It is not a function.

If $4c = 19$ (ignoring the minus sign), $c = \frac{19}{4}$. The focus is $F\left(-\frac{19}{4}, 0\right)$ (seen from the graph); the directrix is $x = +\frac{19}{4}$; $e = 1$ (always); and the vertex is $V(0, 0)$.

Example 43: Describe $(y - 4)^2 = -19(x + 3)$.

Solution: The vertex has been translated to the point $V(-3, 4)$. Again, remember that the x-coordinate goes first. Then c is still $\frac{19}{4}$. $F\left(-3 - \frac{19}{4}, 4\right)$, and the directrix is $x = -3 + \frac{19}{4}$.

Example 44: Describe $3x^2 - 24x + 6y - 9 = 0$.

Solution: First, divide by 3 (the coefficient of the square term) and isolate all terms with the square variable on one side, then complete the square.

$x^2 - 8x = -2y + 3$; $x^2 - 8x + 16 = -2y + 3 + 16$; $(x - 4)^2 = -2y + 19$;
$(x - 4)^2 = -2\left(y - \frac{19}{2}\right)$.
The parabola opens down; $V\left(4, \frac{19}{2}\right)$; $4c = 2$; $c = \frac{1}{2}$; $F\left(4 - \frac{1}{2}, \frac{19}{2}\right)$;
the directrix is $y = 4 + \frac{1}{2}$; $e = 1$.

Curve Sketching: Polynomial Intercepts and Rational Function Intercepts and Asymptotes

The people who create the ACCUPLACER exam hint that curve sketching could be on the test. The ACCUPLACER won't actually ask you to sketch curves (this is a multiple-choice test), but they very well could ask you to find intercepts and asymptotes.

Example 45: Find all intercepts for the polynomial $y = x^2 - 2x - 8$

Solution: Just like for a line, to find the x intercept(s), let y = 0; and to find the y-intercept, let $x = 0$. x-intercept: $y = 0 = x^2 - 2x - 8 = (x - 4)(x + 2)$. So $x = 4$ and $x = -2$, and the x-intercepts are (4, 0) and (−2, 0). y-intercept: let $x = 0$ in the original: $y = 0^2 - 2(0) - 8 = -8$; so the y-intercept = (0, −8)

Note *All polynomials have one y-intercept.*

Example 46: Find the intercepts for $y = x(9 - x)(x - 5)$.

Solution: Set each factor = 0; we get (0, 0), (9, 0), and (5, 0). Since there is only one y-intercept, that is (0, 0). (We love when this happens, it's less work!)

Example 47: $y = (x - 1)^4(x + 2)^2(x - 3)^3$

Solution: x –intercepts: $y = 0$; ignore the exponents; we get (1, 0), (−2, 0), (3, 0)
y: intercept, $x = 0$;
so $y = (-1)^4(2^2)(-3)^3 = -108$. The y-intercept is (0, −108).

Definition: **Degree** of a polynomial in one unknown: the highest power of x.

Example 48: Find the degrees of each of the polynomials:

Solution: Example 45: Degree is 2.

Example 46: The degree is 3.

Example 47: if we were to multiply it out (we wouldn't dream of it!!!!!), the first, highest power term would be x^9; so the degree is 9 ($x^4 \times x^3 \times x^2$).

Rational Functions

Definition: A rational function is a polynomial divided by a polynomial.
Vertical asymptote (for rational functions): vertical asymptotes are lines to which a curve gets "very close," "at the end" (when x gets very big or very small [x goes to $+\infty$ or $-\infty$]) but doesn't hit. They are found by setting the denominator equal to zero.

Note *Since polynomials have no denominators they have no asymptotes.*

Example 49: Let $y = \dfrac{(x-3)(x-7)}{x(x+2)(x-15)}$. Find all the vertical asymptotes.

Solution: Set the denominator equal to 0: we have $x = 15$, $x = -2$, and $x = 0$, (the y-axis).

Example 50: Let $y = \dfrac{1}{(x-1)^5(2x-3)^7}$; find all vertical asymptotes.

Solution: Again we ignore the exponents: $x = 1$ and $x = \dfrac{-3}{2}$.

Horizontal Asymptotes

Case 1: If the degree of the top < degree of the bottom, the horizontal asymptote is $y = 0$, the x-axis.

Example 51: Let $y = \dfrac{4x^2 - 5x}{7 + 9x^3}$. Find the horizontal asymptote.

Solution: Degree of the top is 2; degree of the bottom is 3. The horizontal asymptote is $y = 0$. Why should you believe me? Divide every term top and bottom by x^3. We get $y = \dfrac{\frac{4}{x} - \frac{5}{x^2}}{\frac{7}{x^3} + 9}$. As x gets very big (goes to $+\infty$ or very small (x goes to $-\infty$), all the terms with x in it go to 0. So we have essentially $\frac{0}{9} = 0$.

Case 2: If the degree of the top = the degree of the bottom, the horizontal asymptote is $y = \frac{a}{b}$, where a is the coefficient of the highest power on top and b is the coefficient of the highest power on the bottom.

Example 52: Let $y = \dfrac{3x^2 + 4x}{5 - 7x^2}$.

Solution: Since the degree of the top = the degree of the bottom (= 2), the horizontal asymptote is $y = \frac{3}{-7}$ since 3 is the coefficient of the highest power on top and -7 (remember the minus sign) is the coefficient of the highest power on the bottom. Again why should you believe me? Divide every term by x^2. We get $y = \dfrac{3 + \frac{4}{x}}{\frac{5}{x^2} - 7}$; again the fractions go to zero as x gets very big or very small; we get $y = \frac{3}{-7}$ or $\frac{-3}{7}$.

Let's have a problem with everything in it!

Example 53: Find all the intercepts and asymptotes for $y = \dfrac{(x+1)^{40}(x+3)^{77}}{(x-1)^{71}(x-3)^{74}}$

Solution: x-intercepts, $y = 0$, means numerator = 0: $(-1, 0)$ and $(-3, 0)$.

y-intercept, $x = 0$; $y = \dfrac{(1)^{40}(3)^{77}}{(-1)^{71}(-3)^{74}} = -1(3)^3 = -27$: y intercept is $(0, -27)$ vertical asymptote means the denominator = 0: $x = 1$ and $x = 3$ horizontal asymptote: $y = 0$ since the degree of the top is 117 (40 + 77) < degree of bottom 145 (71 + 74).

Sequences

The technical definition of a sequence (on the ACCUPLACER) is a function on the non-negative integers. In other words we have $f(0)$, $f(1)$, $f(2)$,…$f(n)$,… However, we use a different notation. For $f(0)$ we use s_0; for $f(1)$, we use s_1 and so on.

So the sequence $\{s_n\}$, (n is a subscript), means $s_0, s_1, s_2, s_3, \ldots$

Note *Besides the letter n, the letters i, j, k, and m are commonly used for subscripts.*

Note *In many books, s_1 is the first term of the sequence.*

Note *In many books, we have $\{a_n\}$: $a_0, a_1, a_2, a_3, \ldots$ (also b_n and c_n)*

Example 54: Here is a chart with different members of the sequence

Note *For n = 0, 1, 2, 3,… substitute in the expression for the term*

Example 54:

	s_0	s_1	s_2	s_3	s_4	s_k
$S_n = \frac{1}{(n+1)^2}$	$\frac{1}{1}$	$\frac{1}{4}$	$\frac{1}{9}$	$\frac{1}{16}$	$\frac{1}{25}$	$\frac{1}{(k+1)^2}$
$S_n = n!$	$0! = 1$ (definition)	$1! = 1$	$2! = 2$	$3! = 6$	$4! = 24$	$k!$
$S_n = \frac{(-1)^n}{n+6}$	$\frac{1}{6}$	$\frac{-1}{7}$	$\frac{1}{8}$	$\frac{-1}{9}$	$\frac{1}{10}$	$\frac{(-1)^k}{k+6}$

Note *Using $(-1)^n$ or $(-1)^{n+1}$ is the way we alternate signs: n or n + 1 determines whether the first sign is positive or negative.*

Sometimes we form a sequence using previous terms.

Example 55: Let $s_0 = 10$ and let $s_n = 2s_{n1} - 3$, $n \geq 1$.

Solution: Given: For $n = 0$, $s_0 = 10$. For $n = 1$, $s_1 = 2s_0 - 3 = 2(10) - 3 = 17$. For $n = 2$, $s_2 = 2s_{2-1} - 3 = 2(17) - 3 = 31$. You should do a few more for practice. The next terms are 59, 115, and 227.

Example 56: Let $s_0 = 0$ and $s_1 = 1$. Define $s_n = s_{n1} + s_{n2}$, $n \geq 2$.

Solution: We know for $n = 0$ and $n = 1$. For $n = 2$, we have $s_2 = s_1 + s_0 = 1 + 0 = 1$. For $n = 3$, $s_3 = s_2 + s_1 = 1 + 1 = 2$. For $n = 4$, we have $s_4 = s_3 + s_2 = 2 + 1 = 3$. For $n = 5$, $s_5 = s_4 + s_3 = 3 + 2 = 5$. You should check that the next numbers are 8, 13, 21, and 34. This is a very famous sequence. It is called the **Fibonacci** sequence.

Equations with Square Roots

Finally, the last topic! Although the title of this chapter is "Topics that Don't Fit Anywhere Else," the truth is this topic could fit in a number of places. So let's put it here.

Example 57: Solve for x: $\sqrt{2x-1} = 3$

Square both sides: $2x - 1 = 9$

Solve the Equation: $2x = 10$; $x = 5$

With square roots, you must check to see if the solution(s) is (are) OK. $\sqrt{2(5)-1} = 3$; $\sqrt{9} = 3$

Example 58: Solve for x: $\sqrt{2x+1} + x = 7$

Isolate the square root: $\sqrt{2x+1} = 7 - x$

Square both sides: $2x+1 = 49 - 14x + x^2$

Move all terms to one side with zero on the other. The x^2 term should have a positive coefficient: $0 = x^2 - 16x + 48$.

Factor and get the x values: $(x - 4)(x - 12) = 0$; $x = 4$ and 12.

Substitute back in the original Equation (54): You will find $x = 4$ checks and $x = 12$ does not.

Note *On this test when there is one answer, usually it will check.*

Note *On this test, when there are two answers, one will check and one won't. This is NOT always true, but usually on a test like this it is.*

Note *If it is a multiple choice, it is usually shorter to check the answers without doing the problem.*

Chapter 16 Test

1. $6! - 4! =$
2. $\frac{8!}{7!} =$
3. $P(10,3) =$
4. $C(8,2) =$
5. Evaluate $\begin{vmatrix} 0 & 1 & 2 \\ 3 & 4 & 5 \\ -2 & 0 & -1 \end{vmatrix}$.
6. All children must take history or geography. If 30 kids take geography and 40 kids take history, and 27 take both, how many children are there?
7. You have a choice of 5 entrees, 7 desserts, and 10 drinks. How many different meals are there if you can choose one from each group?

For questions 8–13, use the set {4, 6, 7, 8, 9}; an example of a three-digit number is 552. Assume that repetition of digits is permitted.

8. How many three-digit numbers are there?
9. How many three-digit permutations of these numbers are there?
10. How many three-digit odd numbers greater than 700 are there?
11. How many three-digit even numbers greater than 900 are there?
12. How many three-digit numbers have their last two digits the same?
13. How many numbers are less than 800?

14. How many ways can 5 people stand in a line?
15. How many committees of 3 can be formed from 10 people?

For questions 16–21, let the Universe $U = \{a, b, c, d, e, f, g, h\}$; and sets $A = \{a, c, d, f, g\}$; $B = \{a, b, d, f, h\}$; $C = \{c, e, g\}$.

16. Find $A \cup B$.
17. Find $A \cap B$.
18. Find $B \cap C$.
19. Find the complement of A.
20. Find the complement of the set of vowels.
21. Find the probability that 3 heads will occur from 3 tosses of a fair coin.

Questions 22 and 23 refer to a jar that has 4 red balls and 5 yellow balls. Two balls are chosen.

22. Using replacements, find the probability that 2 yellow balls will be picked.
23. Without replacement, find the probability that 2 red balls will be picked.
24. Solve for x: $\begin{vmatrix} x & -4 \\ x & x+3 \end{vmatrix} = 8$.
25. If the point (a, b) is translated 4 to the left and 7 up, what are the coordinates of the new point?
26. What is the 20th term in the sequence 100, 94, 88, 82,...?
27. Find the sum of the 20 terms in question 26.
28. Find the 6th term in the sequence 256, 128, 64,...
29. Find the sum of the six terms in question 28.
30. A ball is dropped from 100 feet. It bounces only 80 feet, and keeps this same ratio on all successive bounces. It does this forever. How many feet does the ball travel?
31. Find the geometric means of -4 and -9.

For questions 32–44, identify the graph. Answer choices are:

one line	hyperbola	two lines
point	circle	no graph
parabola	none of these answers	ellipse

32. $\frac{x}{4} + \frac{y}{5} = 1$.
33. $3x = 11$.
34. $x + y^2 = 7$.
35. $4x^2 = 4 + 4y^2$.
36. $4x^2 = 4 - 4y^2$.
37. $y^2 = 11$.
38. $4x^2 + 9y^2 = 0$.
39. $4x^2 - 9y^2 = 0$.
40. $(x - 4)^2 + (y - 7)^2 + 11 = 0$.
41. $2x^2 + 3 = 4y^2$.
42. $4y^2 = 3 - 2x^2$.
43. $y = \frac{1}{x}$.
44. $y = \frac{1}{(x-2)^2}$.

Questions 45 and 46 refer to the equation $2x^2 + 2y^2 - 4x + 8y - 12 = 0$.

45. Find the center.
46. Find the diameter.

Questions 47–51 refer to the equation $\frac{(y-9)^2}{16} - \frac{(x+8)^2}{25} = 1$.

47. Find the center.
48. Find the vertices.
49. Find the foci.
50. Find the asymptotes.
51. Find the eccentricity.

Questions 52–56 refer to the equation $\frac{(y-7)^2}{64} + \frac{(x+9)^2}{100} = 1$.

52. Find the center.
53. Find the vertices.
54. Find the foci.
55. Find the asymptotes.
56. Find the eccentricity.

Questions 57–60 refer to the equation $(x-3)^2 = 12y + 48$.

57. Find the vertex.
58. Find the focus.
59. Find the directrix.
60. Find the eccentricity.

Questions 61–63 refer to the equation $y = (x-1)(4-x)(x+5)$.

61. List all the x-intercepts.
62. List all the y-intercepts.
63. What is the degree of this polynomial?

Questions 64–66 refer to the equation $y = x^5(x-2)^3(5+x)$.

64. List all the x-intercepts.
65. List all the y-intercepts.
66. What is the degree of the polynomial?

Questions 67–70 refer to the equation $y = \dfrac{(3x+5)(x+1)}{(x-2)(7-5x)}$.

67. List all the *x*-intercepts.

68. List all the *y*-intercepts.

69. List all the vertical asymptotes.

70. List all the horizontal asymptotes.

Questions 71–74 refer to the equation $y = \dfrac{(x-1)^{40}(x-5)^{53}}{(x+1)^{47}\ (x+5)^{51}}$.

71. List all the *x*-intercepts.

72. List all the *y*-intercepts.

73. List all the vertical asymptotes.

74. List all the horizontal asymptotes.

Questions 75–78 refer to the equation $y = \dfrac{(3x-5)^7(4x+9)^{11}}{(x)^3(2-x)^{13}}$.

75. List all the *x*-intercepts.

76. List all the *y*-intercepts.

77. List all the vertical asymptotes.

78. List all the horizontal asymptotes.

79. If $s_n = (-1)^{n+3}/(n^2+5)$, list the first 5 terms.

80. If $s_0 = 5$ and $s_1 = 7$, find s_2 if $s_n = 4s_{n-2} - 3s_{n-1}$

81. Solve for *x*: $\sqrt{4x+9} = 5$.

82. Solve for *x*: $\sqrt{x+1} + x = 11$.

Answers to Chapter 16 Test

1. $720 - 24 = 696$.
2. 8.
3. $10 \times 9 \times 8 = 720$.
4. $\dfrac{8(7)}{1(2)} = 28$.
5. $(0)(4)(-1) + (1)(5)(-2) + 2(3)(0) - (-2)(4)(2) - (-1)(3)(1) - 0(5)(0) = 9$.

6. $30 + 40 - 27 = 43$.
7. $(5)(7)(10) = 350$ different meals.
8. $5(5)(5) = 125$.
9. $P(5, 3) = 5(4)(3) = 60$.
10. $(3)(5)(2) = 30$.
11. $1(5)(3) = 15$.
12. $(5)(5)(1) = 25$.
13. Read carefully! This includes three-digit numbers, two-digit numbers, and one-digit numbers: $(3)(5)(5) + (5)(5) + 5 = 75 + 25 + 5 = 105$.
14. $5! = 120$.
15. $C(10, 3) = 120$.
16. $\{a, b, c, d, f, g, h\}$.
17. $\{a, d, f\}$.
18. ϕ.
19. $\{b, e, h\}$.
20. $\{b, c, d, f, g, h\}$.
21. $\left(\frac{1}{2}\right)\left(\frac{1}{2}\right)\left(\frac{1}{2}\right) = \frac{1}{8}$.
22. $\left(\frac{5}{9}\right)\left(\frac{5}{9}\right) = \frac{25}{81}$.
23. $\left(\frac{4}{9}\right)\left(\frac{3}{8}\right) = \frac{1}{6}$.
24. $x(x + 3) - (-4)(x) = 8$; $x^2 + 7x - 8$ or $(x + 8)(x - 1) = 0$; $x = -8$ or $x = 1$.
25. $(a - 4, b + 7)$.
26. $a_{20} = a_1 + (n - 1)d$; so $a_{20} = 100 + 19(-6) = -14$.
27. $S = \left(\frac{n}{2}\right)(a_1 + a_{20}) = \frac{20}{2}(100 + (-14)) = 860$.
28. $a_6 = a_1 r^5 = 256\left(\frac{1}{2}\right)^5 = 8$.
29. Start with $256 + 128 + 64 + 32 + 16 + 8 = 504$ or $S = \frac{a(1 - r^n)}{1 - r} = 256\frac{1 - \left(\frac{1}{2}\right)^6}{1 - \left(\frac{1}{2}\right)} = 504$.

30. The first drop gives you 100 feet; then there are two infinite sums (up and down) where the first term is 80 and the ratio is $\frac{4}{5} = \frac{80}{100}$; so we get $100 + \frac{2a}{1-r} = 100 + \frac{2(80)}{1-\frac{4}{5}} =$ 900 feet total distance traveled.
31. $\pm\sqrt{(-4)(-9)} = \pm 6$.
32. One line.
33. One line! Remember this is two dimensions!
34. Parabola.
35. Hyperbola; put the y^2 term on the left.
36. Circle; put the y^2 term on the left.
37. Two lines.
38. One point: (0, 0).
39. Two lines; $y = \pm\frac{2}{3}x$.
40. ϕ: put the number on the right side; the sum of two positives cannot be a negative.
41. Hyperbola; put the y^2 term on the left.
42. Ellipse; put the x^2 term on the left.
43. Hyperbola; $xy = 1$.
44. None of these.
45. $(x-1)^2 + (y+2)^2 = 11$; center is at: $(1, -2)$.
46. $2\sqrt{11}$; the diameter, not the radius!
47. It could be argued that it has no center, but where the asymptotes cross is $(-8, 9)$.
48. $(-8, 9 \pm 4) = (-8, 5)$ and $(-8, 5)$, $(-9, 7 \pm 8) = (-9, -1)$ and $(-9, 15)$
49. $(-8, 9 \pm \sqrt{41})$.
50. $(y-9) = \pm\frac{4}{5}(x+8)$.
51. $\frac{\sqrt{41}}{4}$.
52. $(-9, 7)$.
53. $(-9 \pm 10, 7) = (-19, 7)$ and $(1, 7)$
54. $(-9 \pm 6, 7) = (-15, 7)$ and $(-3, 7)$
55. None; ellipses don't have asymptotes.
56. $e = \frac{6}{10} = \frac{3}{5}$.
57. $(x-3)^2 = 12(y+4)$; $4c = 12$; $c = 3$; $V(3, -4)$.

58. $(3, -4 + 3) = (3, -1)$.

59. $y = -4 - 3$, or $y = -7$.

60. $e = 1$ (always).

61. $(1, 0)$, $(4, 0)$, and $(-5, 0)$, setting each factor $= 0$.

62. Substitute $x = 0$, we get the point $(0, -20)$.

63. Degree is 3 since if you multiplied out the polynomial we get $-x^3$ as the highest power of x.

64. $(0, 0)$, $(2, 0)$, and $(-5, 0)$.

65. $(0, 0)$.

66. Degree is 9.

67. $y = 0$ means the top of fraction $= 0$; $(\frac{-5}{3}, 0)$ and $(-1, 0)$.

68. Let $x = 0$; so $y = \frac{5}{-14}$; y-intercept is $(0, \frac{-5}{14})$.

69. Vertical asymptotes bottom $= 0$: $x = 2$ and $x = \frac{7}{5}$.

70. Horizontal asymptote: degree of top = degree of bottom = 2: $y = \frac{3}{-5} = \frac{-3}{5}$.

71. Top $= 0$: $(1, 0)$ and $(5, 0)$.

72. $x = 0$: $(0, -25)$ since $\frac{(-5)^{53}}{5^{51}} = -25$.

73. Bottom $= 0$: $x = -1$ and $x = -5$.

74. Horizontal asymptote: $y = 0$ since the degree of the top $= 93 <$ degree of bottom $= 98$.

75. $\left(\frac{5}{3}, 0\right)$ and $\left(\frac{-9}{4}, 0\right)$.

76. No y-intercepts since $x = 0$ is in the bottom (vertical asymptote on the y-axis): a rational functional has 0 or 1 y intercepts.

77. $x = 0$ and $x = 2$.

78. No horizontal asymptote, since the degree of the top $= 18 >$ degree of bottom $= 15$. A rational function has 0 or 1 horizontal asymptotes.

79. $\frac{-1}{5}, \frac{1}{6}, \frac{-1}{9}, \frac{1}{14}$, and $\frac{-1}{21}$. Let $n = 0,1,2,3,4$:

80. $s_2 = 4(5) - 3(7) = -1$.

81. $x = 4$.

82. $x = 8$.

Now that you have finished the review, let's practice what you've learned.

Answer to "Bob Asks": \$40. The radius is doubled, but $A = \pi r^2$ means the area is $2^2 = 4$ times larger.

CHAPTER 17: *12 Practice Tests*

Arithmetic Practice Tests

Directions: For each of the Arithmetic practice tests, choose the best answer from the four choices given.

Arithmetic: Test 1

1. 3.45 + .123 + 51.6 =

A. 55.173

B. 56.28

C. 98.4

D. 98.473

2. Reduce $\frac{28}{42}$.

A. $\frac{1}{2}$

B. $\frac{2}{3}$

C. $\frac{3}{4}$

D. $\frac{7}{11}$

3. .3% written as a decimal is

A. 0.0003

B. 0.003

C. 0.03

D. 0.3

4. $\frac{3}{8} + \frac{2}{5} =$

A. $\frac{5}{13}$

B. $\frac{3}{20}$

C. $\frac{15}{16}$

D. $\frac{31}{40}$

5. 5.95×59.7 is slightly less than

A. 36

B. 360

C. 3600

D. 36,000

6. $\frac{3}{5} \div 3 =$

A. $\frac{1}{5}$

B. $\frac{9}{5}$

C. $\frac{19}{5}$

D. $\frac{27}{5}$

7. What is 14% of 14?

A. 1

B. 1.96

C. 100

D. 196

8. An 80-foot board is cut in the ratio of 2 to 3. The longer piece is

A. 16

B. 32

C. 48

D. 64

9. $\frac{3}{8}$ as a decimal to the nearest hundredth is

A. 3.80

B. 3.67

C. 0.33

D. 0.38

10. 3 yards 4 feet 5 inches in inches is

A. 345

B. 161

C. 143

D. 137

11. Three workers do a job. The first one does $\frac{1}{3}$ of it; the second one does $\frac{1}{4}$ of it. In order to finish the job, what part must the third worker do?

A. $\frac{1}{12}$

B. $\frac{5}{12}$

C. $\frac{7}{12}$

D. $\frac{11}{12}$

12. 9% of what number is 18?

A. 1.62

B. 2

C. 162

D. 200

13. $1\frac{3}{4} \times 1\frac{3}{4} =$

A. 1

B. $2\frac{9}{16}$

C. $3\frac{1}{16}$

D. $3\frac{1}{2}$

14. $42 \div .03 =$

A. 1.26

B. 14

C. 126

D. 1400

15. $4 - 1\frac{3}{5} =$

A. $2\frac{2}{5}$

B. $3\frac{2}{5}$

C. $4\frac{3}{5}$

D. $5\frac{3}{5}$

16. A high school basketball team made 70% of its free throws. If the team made 280 free throws, how many free throws did it take?

A. 196

B. 220

C. 300

D. 400

17. How many hours are there in a week?

A. 140

B. 152

C. 160

D. 168

Arithmetic: Test 1 Answers

1. **A:** Line up decimal point and add the numbers together.

2. **B:** $\frac{28}{42} = \frac{2 \times 2 \times 7}{2 \times 3 \times 7} = \frac{2}{3}$ after canceling the sevens and one of the twos top and bottom.

3. **B:** Move the decimal point two places to the left and drop the % sign.

4. **D:** $\frac{15}{40} + \frac{16}{40}$

5. **B:** $(6)(60) = 360$

6. **A:** $\frac{3}{5} \times \frac{1}{3} = \frac{1}{5}$

7. **B:** $14 \times 0.14 = 1.96$

8. **C:** The board is divided into 5 parts $(2 + 3)$; $\frac{80}{5} = 16$; the bigger piece is three parts, and $3(16) = 48$

9. **D:** $\frac{3}{8} = .375$ rounded to .38

10. **B:** $3(36) + 4(12) + 5 = 161$

11. **B:** $\frac{1}{3} + \frac{1}{4} = \frac{7}{12}$; one job $= 1$; $1 - \frac{7}{12} = \frac{5}{12}$

12. D: $\frac{18}{.09} = 200$

13. C: $\left(\frac{7}{4}\right)\left(\frac{7}{4}\right) = \frac{49}{16} = 3\frac{1}{16}$

14. D: $\frac{42}{.03} = \frac{4200}{3} = 1{,}400$

15. A: $3\frac{5}{5} - 1\frac{3}{5} = 2\frac{2}{5}$

16. D: $\frac{280}{.70} = 400$

17. D: $24(7) = 168$

Arithmetic: Test 1 Answer Key

Question #	Answer Choice	Chapter	Subtopic to Review
1	A	2	Decimals
2	B	2	Fractions
3	B	2	Decimals
4	D	2	Fractions
5	B	2	Approximations
6	A	2	Fractions
7	B	2	Percentages
8	C	7	Ratios
9	D	2	Decimals and Fractions
10	B	7	Measurements
11	B	7	Ratios
12	D	2	Percentages
13	C	2	Fractions
14	D	2	Decimals
15	A	2	Fractions
16	D	2	Percentages
17	D	7	Measurements

Arithmetic: Test 2

1. The product of 3.46 and .2 is

A. 3.66

B. 3.26

C. 0.692

D. 17.3

2. Which is the largest?

A. $\frac{3}{4}$

B. $\frac{7}{8}$

C. $\frac{11}{12}$

D. $\frac{15}{16}$

3. $2\frac{3}{5} \div 1\frac{3}{10} =$

A. $\frac{1}{2}$

B. $1\frac{3}{5}$

C. 2

D. $3\frac{19}{50}$

4. A property tax rate is $1.20 per $100 of evaluation. If a house is evaluated for tax purposes at $4,000, what is the property tax?

A. $48

B. $480

C. $4800

D. $48,000

5. Which of the following values is the smallest?

A. $\frac{4}{9}$

B. 0.44

C. $\frac{3}{7}$

D. 400%

6. $4.123 - .789 =$

A. 4.444

B. 3.444

C. 3.344

D. 3.334

7. 3 pounds 4 ounces is how many ounces?

A. 34

B. 40

C. 52

D. 6,004

8. 60% of the children have red hair. If there are 60 children, how many have red hair?

A. 1

B. 6

C. 36

D. 100

9. Reduced, $\frac{42}{58} =$

A. $\frac{2}{4}$

B. $\frac{3}{4}$

C. $\frac{21}{29}$

D. $\frac{23}{24}$

10. $4\frac{2}{3} + 5\frac{3}{4} =$

A. $9\frac{5}{12}$

B. $9\frac{5}{7}$

C. $10\frac{5}{12}$

D. $20\frac{1}{2}$

11. 72% as a fraction is

A. $\frac{7}{2}$

B. $\frac{18}{25}$

C. $\frac{18}{250}$

D. $\frac{18}{2500}$

12. $2.34 + 3.23 + 4.42 =$

A. 999

B. 99.9

C. 9.99

D. 10

13. Three of every 4,000 light bulbs are defective. If 21 are defective, how many light bulbs do we have?

A. 28

B. 280

C. 2800

D. 28,000

14. If 45 of 60 people like a TV show, what percent do *not* like the show?

A. 25%

B. 45%

C. 55%

D. 75%

15. $5\frac{2}{7} - 2\frac{5}{7} =$

A. $2\frac{3}{7}$

B. $2\frac{4}{7}$

C. $2\frac{5}{7}$

D. $3\frac{5}{7}$

16. 9 is what % of 8?

A. 88.9%

B. 112.5%

C. 889%

D. 1125%

17. $3\frac{1}{5} \times 3\frac{1}{5} =$

A. 1

B. $9\frac{1}{25}$

C. $10\frac{6}{25}$

D. $27\frac{1}{25}$

Arithmetic: Test 2 Answers

1. **C:** (3.46)(.2) = 0.692. 2 decimal places plus one is 3 decimal places to the left.

2. **D:** All four fractions are less than one, and the top is one less than the bottom. The largest will always be the fraction with the largest numbers in them.

3. **C:** $\left(\frac{13}{5}\right)\left(\frac{10}{13}\right) = 2$

4. **A:** \$12.00 for \$1,000; so \$48.00 for \$4,000

5. **C:** A = 0.444, B = 0.440, C = 0.429, D = 4.000. All decimal places need to be taken far enough to see which is the smallest (or largest in another problem): although 3 are shown, only 2 are needed here.

6. **D:** Line up the decimal points and subtract.

7. **C:** 3(16) + 4 = 52

8. **C:** (.60)(60) = 36

9. **C:** $\frac{2 \times 3 \times 7}{2 \times 29} = \frac{21}{29}$

10. **C:** $4\frac{8}{12} + 5\frac{9}{12} = 9\frac{17}{12} = 10\frac{5}{12}$

11. **B:** $\frac{72}{100} = \frac{18}{25}$

12. **C:** Line up the decimal points and add the numbers.

13. **D:** 7(4000) = 28,000. 3(7) = 21; so 7(4000) = 28,000 is the total number of light bulbs.

14. **A:** $\frac{45}{60} = \frac{3}{4}$ like it; so $1 - \frac{3}{4} = \frac{1}{4}$ don't. $\frac{1}{4} = 25\%$

15. **B:** $4\frac{9}{7} - 2\frac{5}{7} = 2\frac{4}{7}$

16. **B:** $\frac{9}{8} = 1.125 = 112.5\%$

17. **C:** $\left(\frac{16}{5}\right)\left(\frac{16}{5}\right) = \frac{256}{25} = 10\frac{6}{25}$

Arithmetic: Test 2 Answer Key

Question #	Answer Choice	Chapter	Subtopic to Review
1	C	2	Decimals
2	D	2	Fractions
3	C	2	Fractions
4	A	7	Ratios
5	C	2	Decimals
6	D	2	Decimals
7	C	7	Measurements
8	C	2	Percentages
9	C	2	Fractions
10	C	2	Fractions
11	B	2	Percentages
12	C	2	Decimals
13	D	7	Ratios
14	A	2	Percentages
15	B	2	Fractions
16	B	2	Percentages
17	C	2	Fractions

Arithmetic: Test 3

1. The difference between 1 and 0.237 is

A. 1.237

B. 0.763

C. 0.873

D. 0.237

2. 22% of what number is 55?

A. 12.10

B. 250

C. 1210

D. 25,000

3. Which of the following fractions is largest?

A. $\frac{3}{2}$

B. $\frac{22}{25}$

C. $\frac{19}{20}$

D. $\frac{77}{77}$

4. (.03)(.003) =

A. 0.000009

B. 0.00009

C. 0.0009

D. 0.009

5. $4\frac{1}{2} \times \frac{2}{9} =$

A. 0

B. $\frac{4}{81}$

C. 1

D. $20\frac{1}{4}$

6. The area of a square is 16. The perimeter is

A. 4

B. 8

C. 16

D. 256

7. 2,000 centimeters is how many meters?

A. 20,000

B. 200

C. 20

D. 0.2

8. $5\frac{1}{3} - 3\frac{1}{5} =$

A. $1\frac{2}{15}$

B. 2

C. $2\frac{2}{15}$

D. $8\frac{8}{15}$

9. An \$80 radio is discounted 15%. How much does it cost?

A. \$12

B. \$60

C. \$68

D. \$92

10. 16 is what percent of 20?

A. 4

B. 40

C. 80

D. 125

11. $\frac{1}{4} \div 24 =$

A. 96

B. 6

C. $\frac{1}{6}$

D. $\frac{1}{96}$

12. Two angles of a triangle are the same; the third angle is 8 times one smaller angle. The largest angle of the triangle is how many degrees?

A. 18

B. 38

C. 72

D. 144

13. Which of the following is the largest?

A. 3%

B. $\frac{1}{3}$

C. $\frac{4}{13}$

D. 0.34

14. $4\frac{1}{6} + 7\frac{11}{12} =$

A. $11\frac{1}{12}$

B. $11\frac{12}{18}$

C. $12\frac{1}{12}$

D. $12\frac{11}{12}$

15. $\frac{23.2}{.008} =$

A. 29

B. 290

C. 2900

D. 29,000

16. $16 + 22.7 + 2.34 =$

A. 40.04

B. 41.04

C. 62.1

D. 63.1

17. (1.01)(7.98)(8.02) is approximately equal to

A. 63

B. 64

C. 126

D. 128

Arithmetic: Test 3 Answers

1. **B:** $1 - 0.237 = 0.763$. $1 = 1$. Line up the decimal point.

2. **B:** $\frac{55}{.22} = \frac{5500}{22} = \frac{500}{2}$, dividing top and bottom by 11

3. **A:** It is the only one larger than 1.

4. **B**

5. **C:** $\left(\frac{9}{2}\right)\left(\frac{2}{9}\right) = 1.$

6. **C:** If $4s = 16$, $s = 4$; the area $= 4(4) = 16$.

7. **C:** $\frac{2000}{100} = 20.$

8. **C:** $\frac{16}{3} - \frac{16}{5} = \frac{80}{15} - \frac{48}{15} = \frac{32}{15} = 2\frac{2}{15}$

9. **C:** $\$80 \times .85 = \$68.$

10. **C:** $\frac{16}{20} = .8 = 80\%$

11. **D:** $\left(\frac{1}{4}\right)\left(\frac{1}{24}\right) = \frac{1}{96}.$

12. **D:** 10 units = 180; so one unit is 18; 8 units = 8(18) = 144.

13. **D:** A = .03, B = .33, C = .30, D = .34

14. **C:** $4\frac{2}{12} + 7\frac{11}{12} = 12\frac{1}{12}$

15. **C:** $\frac{23200}{8} = 2900$

16. **B:** Align the decimal points and add the numbers together

17. **B:** $(1)(8)(8) = 64.$

Arithmetic: Test 3 Answer Key

Question #	Answer Choice	Chapter	Subtopic to Review
1	B	2	Decimals
2	B	2	Percentages
3	A	2	Fractions
4	B	2	Decimals
5	C	2	Fractions
6	C	11	Areas
7	C	7	Measurements
8	C	2	Fractions
9	C	2	Percentages
10	C	2	Percentages
11	D	2	Fractions
12	D	10	Angles and Triangles
13	D	2	Decimals, Fractions and Percentages
14	C	2	Fractions
15	C	2	Decimals
16	B	2	Decimals
17	B	2	Approximations

Arithmetic: Test 4

1. The sum of 1 and 0.45 is

A. 0.45

B. 0.55

C. 1

D. 1.45

2. Find the area of the circle whose diameter is 20.

A. 20π

B. 100π

C. 200π

D. 400π

3. $\left(\frac{2}{3}\right)\left(\frac{3}{4}\right)\left(\frac{4}{5}\right) =$

A. $\frac{1}{2}$

B. $\frac{2}{5}$

C. $\frac{5}{6}$

D. 1

4. $20 - \frac{3}{8} =$

A. $19\frac{5}{8}$

B. $20\frac{3}{8}$

C. $20\frac{5}{8}$

D. $53\frac{1}{3}$

5. Find the cost of a $420 TV with a 7% state sales tax.

A. $427

B. $441

C. $449.40

D. $462.50

6. $(.02)(2\%)\left(\frac{1}{50}\right) =$

A. 0.0000008

B. 0.000008

C. 0.00008

D. 0.008

7. What is 10% of 10?

A. 1

B. 10

C. 100

D. 10,000

8. $4\frac{7}{8} + 3\frac{5}{6} =$

A. $7\frac{12}{16}$

B. $7\frac{17}{24}$

C. $8\frac{7}{24}$

D. $8\frac{17}{24}$

9. Find the average (the mean) of $\frac{1}{4}$ and $\frac{1}{8}$.

A. $\frac{1}{6}$

B. $\frac{3}{16}$

C. $\frac{7}{32}$

D. $\frac{11}{64}$

10. Four yards $\frac{1}{2}$ foot is how many inches?

A. 126

B. 150

C. 156

D. 166

11. 60% of the class are girls. The class has 60 students. How many are girls?

A. 36

B. 60

C. 100

D. 1000

12. $3.45 \times 2.1 =$

A. 6.245

B. 7.135

C. 7.145

D. 7.245

13. Round 5.9987 to the nearest tenth.

A. 6.000

B. 6.00

C. 6.0

D. 6

14. The area of a rectangle is 84. The base is 4. The height is

A. 42

B. 38

C. 21

D. 10.5

15. $24\frac{6}{7} \div 6 =$

A. $4\frac{6}{7}$

B. $4\frac{1}{7}$

C. 4

D. $\frac{3}{16}$

16. $\frac{1}{2} + \frac{1}{3} + \frac{1}{4} =$

A. $\frac{1}{24}$

B. $\frac{1}{5}$

C. 1

D. $1\frac{1}{12}$

17. Round 1,234,567 to the nearest 1,000.

A. 1000

B. 1,234,600

C. 1,234,000

D. 1,235,000

Arithmetic: Test 4 Answers

1. **D:** Simply add the numbers together.

2. **B:** $r = 10$; area $= \pi r^2 = 100\pi$.

3. **B:** The 3's and the 4's cancel each other out.

4. **A:** $19\frac{8}{8} - \frac{3}{8} = 19\frac{5}{8}$

5. **C:** $\$420 + (.07)(\$420) = \$449.40$.

6. **B:** $(.02) \times (.02) \times (.02)$

7. **A:** $(.1) \times (10) = 1$

8. **D:** $4\frac{21}{24} + 3\frac{20}{24} = 7 + 1\frac{17}{24} = 8\frac{17}{24}$

9. **B:** $\dfrac{\frac{1}{4} + \frac{1}{8}}{2} = \frac{3}{16}$.

10. **B:** $4(36) + 6 = 150$.

11. **A:** $0.60(60) = 36$.

12. **D:** Simple multiplication.

13. **C:** Since it is to the nearest tenth, the answer is 6.0; the point 0 shows this.

14. **C:** $\frac{84}{4} = 21$.

15. **B:** $\frac{174}{7} \times \frac{1}{6} = \frac{29}{7}$ since $\frac{174}{6} = 29$.

16. **D:** $\frac{6}{12} + \frac{4}{12} + \frac{3}{12} = \frac{13}{12} = 1\frac{1}{12}$

17. **D:** Follow the rules of rounding.

Arithmetic: Test 4 Answer Key

Question #	Answer Choice	Chapter	Subtopic to Review
1	D	2	Decimals
2	B	11	Areas
3	B	2	Fractions
4	A	2	Fractions
5	C	2	Percentages
6	B	2	Decimals, Percentages, and Fractions
7	A	2	Percentages
8	D	2	Fractions
9	B	1	Averages
10	B	7	Measurements
11	A	2	Percentages
12	D	2	Decimals
13	C	2	Approximations
14	C	11	Areas
15	B	2	Fractions
16	D	2	Fractions
17	D	2	Rounding off

Elementary Algebra Practice Tests

Directions: For each of the Elementary Algebra practice tests, choose the best answer from the four choices given.

Elementary Algebra: Test 1

1. $7 - 4 \times 3 - 5(-4) =$

A. 29

B. 15

C. 0

D. -14

2. $\sqrt{6} \times \sqrt{10} =$

A. 60

B. 14

C. 4

D. $2\sqrt{15}$

3. $(4x - 5y)^2 =$

A. $16x^2 + 25y^2$

B. $16x^2 - 25y^2$

C. $16x^2 - 20xy + 25y^2$

D. $16x^2 - 40xy + 25y^2$

4. $2(3x - 7) - (x - 4) = 5.\ x =$

A. 1

B. 3

C. $\frac{23}{6}$

D. $\frac{23}{5}$

5. $\dfrac{x^6x^9y^8}{x^5y^2} =$

A. x^3y^6

B. x^5y^6

C. $x^{10}y^6$

D. xy^4

6. If $x = -3$ and $y = -10$, then $xy^2 - (xy)^2 =$

A. 0

B. -90

C. -469

D. -1200

7. Write in symbols: The product of 4 and the quantity x less 6 is the same as 7 less than x.

A. $4(6 - x) = x - 7$

B. $4(x - 6) = x - 7$

C. $4x - 6 = 7 - x$

D. $4x - 6 = x - 7$

8. $25^{\frac{3}{2}}$

A. 125

B. $37\frac{1}{2}$

C. 5

D. $\sqrt[3]{25^2}$

9. If $x > 100$, $\dfrac{x^2 - 2x - 8}{x^2 - 4x} =$

A. $1 - \dfrac{2}{x}$

B. $\dfrac{x - 4}{x + 2}$

C. $\dfrac{x + 2}{x}$

D. $\dfrac{1}{2} - \dfrac{2}{x}$

10. An apple cost 45 cents and a pear cost 55 cents. The total cost in cents of x apples and y pears is

A. $475xy$

B. $45x + 55y$

C. $45y + 55x$

D. $50(x + y)$

11. $40 - \frac{2}{5}x \geq 36$ is equivalent to

A. $x \geq 10$

B. $x \leq 10$

C. $x \geq 100$

D. $x \leq 100$

12. The volume of a cone is $\frac{1}{3}\pi r^2 h$. If the diameter $d = 10$ and $h = 12$, the volume is

A. 400π

B. 200π

C. 100π

D. 50π

Elementary Algebra: Test 1 Answers

1. **B:** $7 - 12 + 20 = 15$.

2. **D:** $\sqrt{2 \times 2 \times 3 \times 5} = \sqrt{60} = \sqrt{4}\sqrt{15} = 2\sqrt{15}$

3. **D:** $(4x - 5)(4x - 5) = 16x^2 - 20x - 20x + 25$

4. **B:** $6x - 14 - x + 4 = 5; 5x - 10 = 5; 5x = 15; x = 3$

5. **C:** $\frac{x^{15}y^8}{x^5y^2} = x^{10}y^6$. When you divide, you subtract the exponents.

6. **D:** $(-3)(-10)(-10) - ((-3)(-10)(-3)(-10)) = -300 - 900$

7. **B:** "Less" doesn't reverse order but "less than" does.

8. A: The square root of 25, cubed. $\left(\sqrt{25}\right)^3 = 5^3 = 125$

9. C: $\dfrac{(x-4)(x+2)}{x(x-4)} = \dfrac{x+2}{x}$

10. B:

11. B: Multiply by 5; $200 - 2x \geq 180$ or $20 \geq 2x$ or $10 \geq x$ or B

12. C: $\frac{1}{3}\pi(5)(5)12 = 100\pi$ since the radius is half the diameter

Elementary Algebra: Test 1 Answer Key

Question #	Answer Choice	Chapter	Subtopic to Review
1	B	1	Order of Operations
2	D	4	Square Roots
3	D	5	Multiplying Binomials
4	B	6	Solving First Degree Equations
5	C	3	Laws of Exponents
6	D	1	Substitution and Order of Operations
7	B	7	Writing Words as Algebraic Statements
8	A	3	Fractional Exponents
9	C	9	Reducing Algebraic Fractions
10	B	7	Writing Words as Algebraic Statements
11	B	6	Solving Linear Inequalities
12	C	13	Volumes

Elementary Algebra: Test 2

1. $\frac{2x-5}{3} = \frac{5x-2}{7}; x =$

A. -29

B. 29

C. 1

D. -1

2. $4-|-4|-|+7|-\left|-\frac{3}{4}\right| =$

A. $-7\frac{3}{4}$

B. $\frac{1}{4}$

C. $1\frac{3}{4}$

D. $15\frac{3}{4}$

3. Factor completely $4x^2 - 10x + 6$.

A. $(4x - 6)(x - 1)$

B. $2(2x^2 - 5x + 3)$

C. $2(2x - 1)(x - 3)$

D. $2(2x - 3)(x - 1)$

4. Simplify: $4x(x - 3) - 7x(5x - 4)$

A. $-31x^2 - 40$

B. $-31x^2 + 16$

C. $-31x^2 - 40x$

D. $-31x^2 + 16x$

5. Evaluate if $x = -2$ and $y = -3$: $(x + 4)^{y-2}$

A. -32

B. $\frac{1}{32}$

C. 10

D. 32

6. $64x^{12}\left(\frac{2x^{20}}{16x^{16}}\right) =$

A. $8x^{16}$

B. $8x^2$

C. $\frac{3x^2}{2}$

D. $8x^{48}$

7. The area of a trapezoid $A = \frac{1}{2}h\,(b_1 + b_2)$. Suppose the area is 100, $h = 8$, and one base is 7. Find the other base.

A. 14

B. 16

C. 18

D. 20

8. $(2\sqrt{3} + 3\sqrt{2})^2 =$

A. 30

B. $30 + 6\sqrt{6}$

C. $30 + 12\sqrt{6}$

D. 66

9. $4 < 2 - 6x < 10$ is equivalent to

A. $-1 > x > -2.5$

B. $\frac{1}{3} < x < \frac{4}{3}$

C. $-\frac{4}{3} < x < -\frac{1}{3}$

D. $1 < x < 2.5$

10. $\dfrac{-6-(-20)}{(-2)(+7)} =$

A. -1

B. 0

C. 1

D. 24

11. Reduce: $\dfrac{4x^2-9}{4x^2+16x+15}$

A. $\dfrac{-9}{1x+25}$

B. $\dfrac{2x-3}{2x+3}$

C. $\dfrac{2x-3}{2x+5}$

D. Cannot be reduced

12. $\dfrac{6x^{20}+8x^{12}+2x^2}{2x^2} =$

A. $3x^{10}+4x^6$

B. $3x^{10}+4x^6+1$

C. $3x^{18}+4x^{10}$

D. $3x^{18}+4x^{10}+1$

Elementary Algebra: Test 2 Answers

1. **A:** Cross-multiply: $7(2x-5) = 3(5x-2)$ or $14x-35 = 15x-6$ or $-29 = x$ or A

2. **A:** $4-4-7-\dfrac{3}{4} = -7\dfrac{3}{4}$.

3. **D:** A and B are factoring, but only D is complete.

4. **D:** $4x^2-12x-35x^2+28x = -31x^2+16x$.

5. **B:** $2^{-5} = \dfrac{1}{2^5} = \dfrac{1}{32}$

6. A: $\frac{(64x^{12})(2x^{20})}{16x^{16}} = 8x^{16}$ since $\frac{64(2)}{16} = 8$ and $12 + 20 - 16 = 16$.

7. C: $100 = \left(\frac{8}{2}\right)(7 + b_2)$; $25 = 7 + b_2$; $b_2 = 18$. There are other orders that will give you the correct answer.

8. C: Use the FOIL method to get $12 + 12\sqrt{6} + 18 = 30 + 12\sqrt{6}$.

9. C: $2 < -6x < 8$; divide by -6; $-\frac{4}{3} < x < -\frac{1}{3}$. Note the order has changed because we divided by a negative.

10. A: $\frac{14}{-14} = -1$

11. C: $\frac{(2x - 3)(2x + 3)}{(2x + 5)(2x + 3)} = \frac{2x - 3}{2x + 5}$.

12. D: $\frac{6x^{20}}{2x^2} + \frac{8x^{12}}{2x^2} + \frac{2x^2}{2x^2} = 3x^{18} + 4x^{10} + 1$. Divide the coefficients and subtract the exponents.

Elementary Algebra: Test 2 Answer Key

Question #	Answer Choice	Chapter	Subtopic to Review
1	A	6	Linear Equations
2	A	6	Absolute Value
3	D	5	Trinomial Factoring
4	D	5	Simplifying Algebraic Expressions
5	B	1 and 3	Evaluation and Exponents
6	A	3	Exponents
7	C	11	Area of Trapezoid
8	C	4	Square Roots
9	C	6	Linear Inequalities
10	A	1	Order of Operations
11	C	9	Reducing Algebraic Fractions
12	D	3	Exponents

Elementary Algebra: Test 3

1. Solve for x: $0 = 5x + 7 + 9 + 11x$

A. -1

B. 0

C. 1

D. 2

2. Factor $3x^3 - 27x$ completely.

A. $3x(x^2 - 9)$

B. $3x(x - 3)^2$

C. $(3x - 3)(x + 3)$

D. $3x(x - 3)(x + 3)$

3. $2\sqrt{8} \times 3\sqrt{6} =$

A. $6\sqrt{14}$

B. $12\sqrt{18}$

C. $24\sqrt{3}$

D. 72

4. Children's tickets C cost \$5 each; senior tickets S cost \$7; and adult tickets A cost \$10. In order for the concert to make a profit, the amount sold must be at least \$10,000. This expression can be written as:

A. $350\ CSA > 10{,}000$

B. $350\ CSA \geq 10{,}000$

C. $5C + 7S + 10A > 10{,}000$

D. $5C + 7S + 10A \geq 10{,}000$

5.

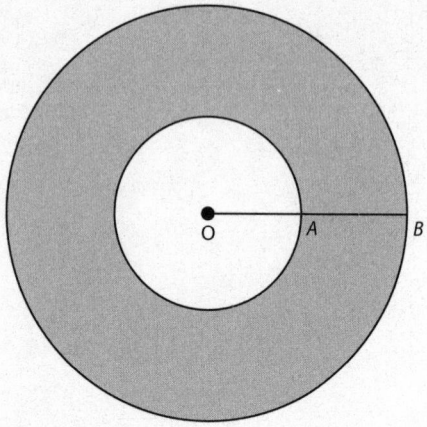

The figure shows two concentric circles (circles having the same center), with center O. $OA = 7$ and $AB = 4$. The area of the shaded portion is

A. 121π

B. 72π

C. 49π

D. 16π

6. $(abc - 6)^2 =$

A. $a^2b^2c^2 - 36$

B. $a^2b^2c^2 + 36$

C. $a^2b^2c^2 + 36abc + 36$

D. $a^2b^2c^2 - 12abc + 36$

7. $\dfrac{-4^2 - 8 \times 4}{-4} =$

A. 24

B. 12

C. 4

D. -8

8. John does $\frac{1}{7}$ of a job; Bill does $\frac{1}{6}$ of a job; and Sam does twice as much as Bill and John together. How much of the job remains to be done?

A. $\frac{1}{14}$

B. $\frac{1}{7}$

C. $\frac{1}{6}$

D. $\frac{1}{3}$

9. $(x^{4a+7})(x^{5a+9}) =$

A. $x^{20a^2 + 71a + 63}$

B. $x^{20a^2 + 63}$

C. $x^{9a + 16}$

D. x^{126a^2}

10. Which of the following is equivalent to $9 - \frac{1}{2}x < 4$?

A. $x > 5$

B. $x < 5$

C. $x > 10$

D. $x < 10$

11. If $x = -1$ and $y = -10$, the value of $3x^2 - 4xy - 2y^2$ is

A. -437

B. -237

C. -157

D. 243

12. $\dfrac{9 - x^2}{(x - 3)^2} =$

A. $\dfrac{x + 3}{x - 3}$

B. $\dfrac{x + 3}{3 - x}$

C. $\dfrac{-1}{6x}$

D. 1

Elementary Algebra: Test 3 Answers

1. C: $0 = 16x + 16$ or $-16x = 16$ or $x = 1$

2. D: Factor out $3x$ to get $3x(x^2 - 9)$. Then $x^2 - 9 = (x - 3)(x + 3)$

3. C: $2\times3\times\sqrt{2\times2\times2\times2\times3} = 6\sqrt{48} = 6\sqrt{16}\sqrt{3} = 24\sqrt{3}$

4. D: "At least" means $\geq$.

5. B: $\pi r^2_{\text{outside}} - \pi r^2_{\text{inside}} = \pi(11^2 - 7^2) = 72\pi$.

6. D: We FOIL; $(abc - 6)(abc - 6) = a^2b^2c^2 - 6abc - 6abc + 36$ or D

7. B: $\dfrac{-16 - 32}{-4} = 12$.

8. A: Sam does $2\left(\dfrac{1}{7} + \dfrac{1}{6}\right) = 2\left(\dfrac{13}{42}\right) = \dfrac{26}{42}$ of the job.

Amount remaining $= 1 - \dfrac{26}{42} - \dfrac{1}{7} - \dfrac{1}{6} = \dfrac{1}{14}$

9. C: Combine like terms to add exponents

10. C: $18 - x < 8$ or $-x < -10$ or $x > 10$. There are other correct ways to do this problem.

11. B: $3(-1)(-1) - 4(-1)(-10) - 2(-10)(-10) = -237$.

12. B: $\dfrac{(3 + x)(3 - x)}{(x - 3)(x - 3)} = -\dfrac{(3 + x)}{(x - 3)} = \dfrac{x + 3}{3 - x} \cdot \dfrac{(3 - x)}{(x - 3)} = -1$ since an expression divided by its negative is always -1.

Elementary Algebra: Test 3 Answer Key

Question #	Answer Choice	Chapter	Subtopic to Review
1	C	6	Linear Equations
2	D	5	Factoring
3	C	4	Square Roots
4	D	6 and 7	Inequalities
5	B	12	Circles
6	D	5	Simplifying Algebraic Expressions
7	B	1 and 3	Order of Operations and Exponents
8	A	7	Problems with Words
9	C	3	Exponents
10	C	6	Linear Inequalities
11	B	1	Substitution and Order of Operations
12	B	5	Factoring and Simplifying Algebraic Fractions

Elementary Algebra: Test 4

1. $2x(3x - 4)(5x + 6) =$

A. $30x^3 - 2x - 24x$

B. $30x^3 - 4x^2 - 48x$

C. $30x^3 + 4x^2 - 48x$

D. $30x^3 - 76x - 48x$

2. Kim bought x apples at 55 cents each and y pears at 45 cents each. Her change in cents from a ten dollar bill can be expressed as

A. $1000 - 55x + 45y$

B. $1000 - 55x - 45y$

C. $1000 - (55x - 45y)$

D. $1000 - 2425xy$

3.

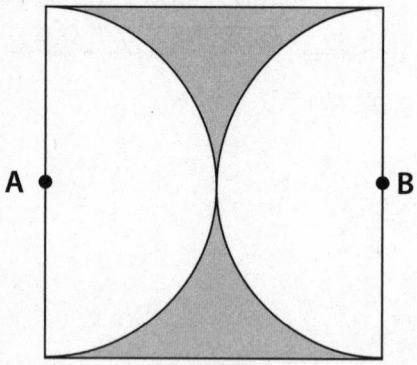

$AB = 1$. The area of the shaded portion of this figure is

A. $4 - \pi$

B. $4 - \frac{\pi}{4}$

C. $1 - \pi$

D. $1 - \frac{\pi}{4}$

4. The area of an equilateral triangle is given by $\frac{s^2\sqrt{3}}{4}$. If the area is $100\sqrt{3}$, then the side $s =$

A. 10

B. 20

C. 40

D. $10\sqrt{3}$

5. $(5ab^3c^9)^3 =$

A. $15ab^9c^{27}$

B. $125ab^9c^{27}$

C. $125a^3b^9c^{27}$

D. $125ab^{27}c^{729}$

6. Solve for x: $\frac{2x+4}{5} = \frac{5x+3}{4}$

A. 17

B. -17

C. $\frac{1}{17}$

D. $-\frac{1}{17}$

7. $V = \pi r^2 h$ is the volume of the cylinder; if the height is tripled and the radius is multiplied by 4, the volume is multiplied by

A. 7

B. 12

C. 48

D. 144

8. If $ax + b + c = p$, $x =$

A. $p - b - c - a$

B. $p - b - \frac{c}{a}$

C. $\frac{p - b - c}{a}$

D. $p - b - c + a$

9. $3^3 + 3^2 + 3 =$

A. 34

B. $2^5 + 2^2$

C. $5^2 + 4^2 - 2$

D. $7^2 - 5^2 + 14$

10. $2\sqrt{27} + 3\sqrt{12} =$

A. $5\sqrt{39}$

B. $12\sqrt{3}$

C. $6\sqrt{324}$

D. 36

11. $1 - 7^2 =$

A. 36

B. -36

C. -48

D. -49

12. $4(2x - 7) - 2(4x - 14) =$

A. 0

B. $-64x^2$

C. -56

D. $-64x^2 - 56$

Elementary Algebra: Test 4 Answers

1. B: $2x(15x^2 - 2x - 24) = 30x^3 - 4x^2 - 48x$

2. B: Notice everything is in cents; so ten dollars is 1000 cents.

3. D: $r = \frac{1}{2}$; area of square − area of circle $= 1 - \pi\left(\frac{1}{2}\right)^2 = 1 - \frac{\pi}{4}$.

4. B: $s^2 = 400$; $s = 20$. Notice the square root of 3 cancels out of the problem.

5. C: $5^3a^3b^{3\times3}c^{9\times3}$

6. C: $4(2x + 4) = 5(5x + 3)$ or $8x + 16 = 25x + 15$ or $17x = 1$ or C

7. C: r^2 is multiplied by 16, and h by 3; $16 \times 3 = 48$.

8. C: $ax = p - b - c$; divide both sides by a.

9. C: Both equal 39.

10. B: $2(3)\sqrt{3}+3(2)\sqrt{3}=6\sqrt{3}+6\sqrt{3}=12\sqrt{3}$

11. C: $1 - 49 = -48$.

12. A: $8x - 28 - 8x + 28 = 0 + 0 = 0$

Elementary Algebra: Test 4 Answer Key

Question #	Answer Choice	Chapter	Subtopic to Review
1	B	5	Algebraic Multiplication
2	B	6	Writing Out in Symbols
3	D	12	Circles
4	B	10	Area of Equilateral Triangle
5	C	3	Exponents
6	C	6	Linear Equations
7	C	13	Volumes
8	C	6	Linear Equations
9	C	1 and 3	Order of Operations and Exponents
10	B	4	Square Roots
11	C	1 and 3	Order of Operations and Exponents
12	A	5	Simplifying Algebraic Expressions

College-Level Mathematics Practice Tests

Directions: For each of the College-Level Mathematics practice tests, choose the best answer from the five choices given.

College-Level Mathematics: Test 1

1. Solve for x: $2x^2 - 5x = 9$; $x =$

A. $\dfrac{-5 \pm \sqrt{-65}}{2}$

B. $\dfrac{-5 \pm \sqrt{97}}{2}$

C. $\dfrac{5 \pm \sqrt{97}}{2}$

D. $\dfrac{5 \pm \sqrt{97}}{4}$

E. $\dfrac{5 \pm \sqrt{-64}}{4}$

2. Cos 120° =

A. $\dfrac{1}{2}$

B. $-\dfrac{1}{2}$

C. $\dfrac{\sqrt{3}}{2}$

D. $-\dfrac{\sqrt{3}}{2}$

E. $\dfrac{\sqrt{2}}{2}$

3. $\text{Log}_{64}x = -\frac{2}{3}; x =$

A. $\frac{1}{16}$

B. -16

C. $\frac{1}{512}$

D. -512

E. 512

4.

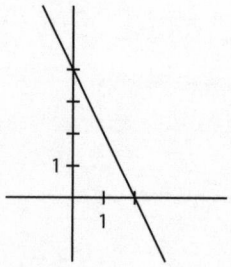

The equation of this line is

A. $y = 2x + 4$

B. $y = 4x + 2$

C. $y = \frac{1}{2}x + 4$

D. $y = -\frac{1}{2}x + 4$

E. $y = -2x + 4$

5. $3^{\frac{8}{3}} - 3^{\frac{5}{3}} - 3^{\frac{5}{3}} =$

A. 3^2

B. $3^{\frac{5}{3}}$

C. 3

D. $3^{\frac{2}{3}}$

E. $3^{-\frac{5}{3}}$

6. If $5x^2 - 7x - 6 = 0$; then $\left(x - \frac{7}{10}\right)^2 =$

A. $\frac{6}{5}$

B. $\frac{6}{10}$

C. $\frac{43}{100}$

D. $\frac{169}{100}$

E. $\frac{121}{100}$

7. If 3 pears and 4 apples cost \$3.55 and 5 pears and 3 apples cost \$3.90, one pear in cents costs

A. 45

B. 50

C. 55

D. 60

E. 65

8. $\csc^2 x - \cot^2 x - \sin^2 x =$

A. $\cos^2 x$

B. $\tan^2 x$

C. $\sec^2 x$

D. $\csc^2 x$

E. $\sin^2 x$

9. Solve for x: $x^2 - 7x + 10 = 4$; $x =$

A. 2 or 5

B. 4 or 3

C. 6 or 1

D. 8 or -1

E. 9 or -2

10. If $\log_3 x - \log_3 (x - 3) = 3; x =$

A. $\frac{81}{26}$

B. $\frac{24}{7}$

C. 9

D. 4

E. No solution

11. $\frac{\sqrt{x}}{\sqrt{x} + \sqrt{y}} =$

A. $\frac{x + \sqrt{xy}}{x + y}$

B. $\frac{x + xy}{x + y}$

C. $\frac{x - \sqrt{xy}}{x - y}$

D. $\frac{x - xy}{x - y}$

E. $\frac{1}{\sqrt{y}}$

12. One vertex of $\frac{(x - 30)^2}{100} + \frac{(y + 20)^2}{64} = 1$ is

A. (36, −20)

B. (24, −20)

C. (30, −14)

D. (30, −26)

E. (20, −20)

13. $\frac{3}{x+4} + \frac{5}{6x+7} =$

A. $\frac{8}{7x+11}$

B. $\frac{8}{(x+4)(6x+7)}$

C. $\frac{23x}{(x+4)(6x+7)}$

D. $\frac{23x+41}{(x+4)(6x+7)}$

E. $\frac{23x+15}{(x+4)(6x+7)}$

14. $\sin^{-1}\left(-\frac{1}{2}\right) =$

A. 30°

B. 60°

C. 150°

D. −30°

E. −60°

15. $g(x) = 3x + 4$ and $f(x) = 2x + 5$; $g(f(x)) =$

A. $6x^2 + 26x + 20$

B. $6x + 13$

C. $6x + 19$

D. $6x^2 + 20$

E. $6x + 25$

16. $4^{3\log_4 5} =$

A. 3

B. 5

C. 15

D. 25

E. 125

17. $\dfrac{x^{-2} - y^{-2}}{x^{-1} - y^{-1}} =$

A. $(x + y)$

B. $\dfrac{xy}{x + y}$

C. $\dfrac{x + y}{xy}$

D. $\dfrac{x - y}{xy}$

E. $\dfrac{xy}{x - y}$

18. $F(x) = 3x - 5; F^{-1}(x) =$

A. $5 - 3x$

B. $\dfrac{1}{3x - 5}$

C. $\dfrac{x + 5}{3}$

D. $x - \dfrac{5}{3}$

E. $x + \dfrac{5}{3}$

19. $\begin{vmatrix} 4 & 5 \\ 6 & -7 \end{vmatrix} =$

A. -840

B. -58

C. -2

D. $+2$

E. $+58$

20. Solve for x: $\sqrt{2x-1} = 7$

A. 4

B. $\frac{15}{2}$

C. 24

D. 25

E. $\frac{97}{4}$

College-Level Mathematics: Test 1 Answers

1. **D:** $a = 2, b = -5, c = -9; x = \frac{-b \pm \sqrt{b^2 - 4ac}}{2a} = \frac{-(-5) \pm \sqrt{(-5)^2 - 4(2)(-9)}}{2(2)}$

$= \frac{5 \pm \sqrt{25+72}}{4} = \frac{5 \pm \sqrt{97}}{4}$

2. **B:** Draw the angle. It is a 60° angle in the second quadrant; $x = -1$, $y = \sqrt{3}$, and $r = 2$; $\cos 120° = \frac{x}{r} = \frac{-1}{2}$

3. **A:** $\log_b x = y$ if $b^y = x$; so $b = 64$ and $y = \frac{-2}{3}$; so we get $64^{-\frac{2}{3}} = \frac{1}{64^{\frac{2}{3}}} = \frac{1}{(\sqrt[3]{64})^2} = \frac{1}{4^2} = \frac{1}{16}$.

4. **E:** The y-intercept is (0, 4) and using the intercepts (0, 4) and (2, 0), $m = \frac{(0-4)}{(2-0)} = -2$

5. **B:** This is a factoring question. Factor out the power of 3 to the lowest exponent. We get $3^{\frac{5}{3}}(3 - 1 - 1) = 3^{\frac{5}{3}}$

6. **D:** $(5x + 3)(x - 2) = 0$, so $x = -\frac{3}{5}$ or $x = 2$. Then $\left(-\frac{3}{5} - \frac{7}{10}\right)^2 = \left(2 - \frac{7}{10}\right)^2 = \frac{169}{100}$.

7. **A:** Let p stand for pears and a stand for apples. $3p + 4a = 3.55$ and $5p + 3a = 3.90$; $3(3p + 4a) - 4(5p + 3a) = 3(3.55) - 4(3.90)$; $-11p = -4.95$; $p = \frac{495}{11}$ (the minus signs cancel)

8. **A:** $(1+\cot^2 x) - \cot^2 x - (1 - \cos^2 x) = \cos^2 x$.

9. **C:** Subtract 4 from each side and then factor. $x^2 - 7x + 6 = 0$ or $(x - 6)(x - 1) = 0$, $x = 6$ or $x = 1$

10. **A:** We use the fact that $\log c - \log d = \log\left(\frac{c}{d}\right)$; $c = x$ and $d = x-3$; we get $\log_3 \frac{x}{x-3} = 3$ or $3^3 = 27 = \frac{x}{x-3}$. Solving we get $27(x-3) = x$; $27x - 81 = x$; $26x = 81$ and $x = \frac{81}{26}$.

11. **C:** $\frac{\sqrt{x}}{\sqrt{x}+\sqrt{x}} \times \frac{\sqrt{x}-\sqrt{y}}{\sqrt{x}-\sqrt{y}} = \frac{x-\sqrt{xy}}{x-y}$

12. **E:** The center is (30, −20); semi major axis is 10; the major vertices are (30 ± 10, −20); the semi minor axis is 8; the minor vertices are (30, −20 ± 8); E is the only one that is here. The semi major axis is the square root of the large denominator; so $10 = \sqrt{100}$. That makes the semi minor axis $\sqrt{64} = 8$

13. **D:** The LCD is $(x+4)(6x+7)$; $\frac{3}{x+4} \times \frac{6x+7}{6x+7} + \frac{5}{6x+7} \times \frac{x+4}{x+4} =$

$$\left(\frac{3}{x+4}\right)\left(\frac{6x+7}{6x+7}\right) + \left(\frac{5}{6x+7}\right)\left(\frac{x+4}{x+4}\right) = \frac{18x+21+5x+20}{(x+4)(6x+7)} = \frac{23x+41}{(x+4)(6x+7)}$$

14. **D:** The answer is in the 4th quadrant; the arc sin is negative. The sine of 30° is $\frac{1}{2}$.

15. **C:** $g(f(x)) = 3(f(x)) + 4 = 3(2x+5) + 4 = 6x + 19$

16. **E:** $4^{3\log_4 5} = 4^{\log_4 5^3} = 4^{\log_4 125}$; by law 8 of logs, the answer is 125.

17. **C:** $\dfrac{\frac{1}{x^2} - \frac{1}{y^2}}{\frac{1}{x} - \frac{1}{y}} \times \frac{x^2y^2}{x^2y^2} = \frac{y^2 - x^2}{xy^2 - x^2y} = \frac{(y-x)(y+x)}{xy(y-x)} = \frac{x+y}{xy}$

18. **C:** Let $y = F(x)$; so $y = 3x - 5$ or $3x = y + 5$ or $x = \frac{(y+5)}{3}$; change the y to x and $x = F^{-1}(x)$ (notation change)

19. **B:** $+(4)(-7) - 6(5) = -58$

20. **D:** Squaring both sides, we get $2x - 1 = 49$; $2x = 50$; $x = 25$; checking back we see 25 gives us an answer. Whenever we have square roots, we must check to see if we have extraneous solutions. On this test you probably don't have to check back.

College-Level Mathematics: Test 1 Answer Key

Question #	Answer Choice	Chapter	Subtopic to Review
1	D	6	Quadratic Formula
2	B	15	Trig
3	A	3 and 14	Logs
4	E	9	Lines
5	B	5 or 6	Factoring or Quadratic Formula
6	D	6 and 1	Quadratic Formula and Substitution
7	A	8 and 2	Equations in 2 Unknowns
8	A	15	Trig
9	C	6	Solving Quadratic Equations by Factoring or the Quadratic Formula
10	A	14	Logs
11	C	4	Square Roots
12	E	16	Ellipses
13	D	5	Adding Algebraic Fractions
14	D	15	Inverse Trig Functions
15	C	14	Functions
16	E	14	Logs
17	C	3 and 5	Exponents and Simplifying Complex Fractions
18	C	14	Inverse Functions
19	B	16	Determinants
20	D	16	Solving Equations with Square Roots

College-Level Mathematics: Test 2

1. $x^4 - 16 = 0; x =$

A. ± 2 only

B. $\pm 2, \pm 2i$ only

C. $\pm 2i$ only

D. 2 and $2i$ only

E. -2 and $-2i$ only

2. Simplify $\sqrt[3]{72x^{32}}$.

A. $6x^{16}\sqrt[3]{2}$

B. $x^{10}\sqrt[3]{72x^2}$

C. $2x^{10}\sqrt[3]{3x^2}$

D. $2x^{10}\sqrt[3]{9x^2}$

E. $4x^{10}\sqrt[3]{9x^2}$

3. $(x + 1)^3 =$

A. $x^3 + 1$

B. $x^3 + x^2 + x + 1$

C. $x^3 + 3x^2 + x + 1$

D. $x^3 + x^2 + 3x + 1$

E. $x^3 + 3x^2 + 3x + 1$

4. $\sin\left[\cos^{-1}\left(\frac{b}{a}\right)\right] =$

A. $\frac{a}{\sqrt{a^2 + b^2}}$

B. $\frac{b}{\sqrt{a^2 + b^2}}$

C. $\frac{\sqrt{b^2 - a^2}}{b}$

D. $\frac{a}{\sqrt{b^2 - a^2}}$

E. $\frac{b}{\sqrt{b^2 - a^2}}$

5. $\log_2(x - 2) + \log_2(x + 4) = 4; x =$

A. 4

B. 6

C. 8

D. 12

E. More than one answer

6. $\frac{5}{6ab^3} - \frac{10}{8a^4b} =$

A. $\frac{-5}{24a^5b^4}$

B. $\frac{-5}{24a^4b^3}$

C. $\frac{-50(4a^3 - 3b^2)}{24a^4b^3}$

D. $\frac{20a^3 - 30b^2}{24a^4b^3}$

E. $\frac{15b^2 - 40a^3}{24a^4b^3}$

7. $\tan \pi =$

A. 0

B. 1

C. -1

D. undefined

E. $\frac{1}{2}$

8. Given: $\frac{x^2}{9} - \frac{y^2}{16} = 1$; the asymptotes are

A. $y = \pm\frac{3}{4}x$

B. $y = \pm\frac{4}{3}x$

C. $y = \pm\frac{3}{5}x$

D. $y = \pm\frac{5}{3}x$

E. $y = \pm\frac{5}{4}x$

9. The first term of an arithmetic progression is 77 and the difference in the terms is -5; the 100th term is

A. -577

B. -423

C. -418

D. -413

E. -408

10. What is $C(7, 4)$?

A. 840

B. 420

C. 140

D. 70

E. 35

11. What line has the same slope as $y = 2x + 3$ and goes through the point $(5, -1)$?

A. $y - 2x = -11$

B. $2y - x = -7$

C. $x = \frac{1}{2}y$

D. $y = 2x^2 - 51$

E. $xy = -5$

12. $\frac{5}{x-3} + \frac{2}{x^2 - 2x - 3} =$

A. $\frac{7}{x^2 - 2x - 3}$

B. $\frac{5x + 7}{x^2 - 2x - 3}$

C. $\frac{7x + 5}{x^2 - 2x - 3}$

D. $\frac{4x + 7}{x^2 - 2x - 3}$

E. $\frac{6x + 4}{x^2 - 2x - 3}$

13. The point $(7, 8)$ translated down 10 and to the right 15 becomes

A. $(-3, 23)$

B. $(17, -7)$

C. $(22, -2)$

D. $(-8, 18)$

E. $(17, 23)$

14. The common point (x, y) of $3x - 5y = 1$ and $y = 5x - 9$ is

A. $(1, 4)$

B. $(2, 1)$

C. $(3, -2)$

D. $(4, 11)$

E. $\left(5, \frac{26}{3}\right)$

15. If $2(10^{2x-3}) = 40$, $\log 2 = .30$, and $\log 3 = .48$, then $x =$

A. 1.35

B. 2.15

C. 3.15

D. 4.75

E. 8.25

16. If $f(x) = \dfrac{3}{x-5}$, then $f^{-1}(x) =$

A. $\dfrac{x-5}{3}$

B. $\dfrac{5x+3}{x}$

C. $\dfrac{x}{5x+3}$

D. $\dfrac{5x-3}{x}$

E. $\dfrac{x}{5x-3}$

17. $\dfrac{7-i}{5-2i} =$

A. $\dfrac{37}{29} + i\dfrac{9}{29}$

B. $\dfrac{33}{29} + i\dfrac{19}{29}$

C. $\dfrac{37}{21} + i\dfrac{9}{21}$

D. $\dfrac{33}{21} + i\dfrac{19}{21}$

E. $\dfrac{49}{21} + i\dfrac{1}{21}$

18. If $(x - 4)(x - 8) < 12$, then

A. $4 < x < 8$

B. $x < 4$ or $x > 8$

C. $2 < x < 10$

D. $x < 2$ or $x > 10$

E. $5 < x < 7$

19. $\sin(x + 180°)$ always equals

A. $\cos x$

B. $-\cos x$

C. $\sin x$

D. $-\sin x$

E. $\sin(90° - x)$

20. A lottery has a player choosing one powerball number from 1 to 40 and five numbers from 1 to 56. How many different combinations are possible?

A. $C(96, 6)$

B. $P(96, 6)$

C. $40 \times P(56, 5)$

D. $40 \times C(56, 5)$

E. $P(40, 5) \times C(56, 5)$

College-Level Mathematics: Test 2 Answers

1. **B:** Factors are $\left(x^2-4\right)\left(x^2+4\right)=\left(x-2\right)\left(x+2\right)\left(x-2i\right)\left(x+2i\right)$.

2. **D:** $\sqrt[3]{2 \times 2 \times 2 \times 3 \times 3} = 2\sqrt[3]{9}; \sqrt[3]{x^{32}}$, 3 into 32 is 10, remainder 2 or $x^{10}\sqrt[3]{x^2}$

3. **E:** $(x + 1)(x + 1) = x^2 + 2x + 1; (x + 1)(x^2 + 2x + 1) = x^3 + 2x^2 + x + x^2 + 2x + 1 = x^3+3x^2+3x+3$

4. **C:** $\cos\theta = \frac{a}{b} = \frac{x}{r}; x = a$ and $r = b; y = \sqrt{b^2 - a^2}; \sin\theta = \frac{y}{r} = C = \frac{\sqrt{b^2-a^2}}{b}$

5. A: Substitution is the easiest way. the long way is as follows: $\log_2(x-2)(x+4) = 4$; so $(x-2)(x+4) = 2^4 = 16$; simplifying we get $x^2 + 2x - 8 = 16$ or $x^2 + 2x - 24 = (x+6)(x - 4) = 0$; $x = -6$ does not check; $x = 4$.

6. D: $\frac{5}{6ab^3}\left(\frac{4a^3}{4a^3}\right) - \frac{10}{8a^4b}\left(\frac{3b^2}{3b^2}\right) = \frac{20a^3 - 30b^2}{24a^4b^3}$

7. A: The point on the x-axis we choose is $(-1, 0)$; $\tan = \frac{y}{x} = \frac{0}{-1} = 0$

8. B: $y = \pm\frac{\sqrt{\text{ of the term under } y^2}}{\sqrt{\text{ of the term under } x^2}}x = \pm\frac{\sqrt{16}}{\sqrt{9}}x = \pm\frac{4}{3}x$

9. C: $l = a + (n - 1)d = 77 + 99(-5) = -418$

10. E: $C(7, 4) = \frac{7 \times 6 \times 5 \times 4}{4 \times 3 \times 2 \times 1} = 35.$

11. A: $y = mx + b$; $m = 2$, $x = 5$, and $y = -1$; so $b = -11$.

12. B: $\frac{5}{x-3} \times \frac{x+1}{x+1} + \frac{2}{(x-3)(x+1)} = \frac{5x+7}{x^2 - 2x - 3}.$

13. C: $(7 + 15, 8 - 10) = (22, -2)$.

14. B: $3x - 5(5x - 9) = 1$; $-22x = -44$; $x = 2$; so $y = 5x - 9 = 5(2) - 1 = 1$.

15. B: $10^{2x-3} = 20$; log = $\log_{10}$ of both sides; $(2x - 3)\log 10 = \log 2 + \log 10$; $(2x - 3)1 = 1.30$; $2x = 4.30$; $x = 2.15$.

16. B: Let $y = f(x)$; $y = \frac{3}{(x-5)}$; $xy - 5y = 3$; $x = \frac{(5y+3)}{y}$; let $y = x$ (notation change) and x is $f^{-1}(x)$.

17. A: Multiply top and bottom by $5 + 2i$. $\frac{7-i}{5-2i}\ \frac{5+2i}{5+2i} = \frac{35 + 9i - 2i^2}{25 - 4i^2} = \frac{35 + 9i + 2}{25 + 4} = \frac{37 + 9i}{29}$

18. C: $x^2 - 12x + 20 = (x - 2)(x - 10) < 0$.

19. D: Adding 180° changes the sign of the sign; sin in the first two quadrants is positive; the sine in the third and fourth quadrants is negative.

20. D: Although it is the smallest number, the answer is D. But it is not small; the answer by calculator is 152, 792, 640; you probably have a better chance being hit by lightning 10 times in one storm than winning, but if you don't get too carried away, it is harmless fun.

College-Level Mathematics: Test 2 Answer Key

Question #	Answer Choice	Chapter	Subtopic to Review
1	B	5	Solving Equations by Factoring
2	D	4	Cube Roots
3	E	5	Simplifying (Multiplying Out an Algebraic Expression)
4	C	15	Inverse Trig Functions
5	A	14	Logs
6	D	5	Adding (Subtracting) Algebraic Fractions
7	A	15	Trig
8	B	16	Hyperbolas
9	C	16	Arithmetic Progression
10	E	16	Combinations
11	A	9	Lines
12	B	5	Adding (Subtracting) Algebraic Fractions
13	C	16	Translations
14	B	8	Solving 2 Equations in 2 Unknowns
15	B	14	Logs
16	B	14	Inverse Functions
17	A	4	Complex Numbers
18	C	6	Quadratic Inequalities
19	D	15	Trig
20	D	16	Permutations and Combinations

College-Level Mathematics: Test 3

1. $\dfrac{3x+2}{x^2-4}+\dfrac{2x+3}{x^2-4x+4}=$

A. $\dfrac{5x+5}{2x^2-4x}$

B. $\dfrac{5x+5}{(x-2)^2(x+2)}$

C. $\dfrac{5x^2+3x+2}{(x-2)^2(x+2)}$

D. $\dfrac{5x^2-3x+2}{(x-2)^2(x+2)}$

E. $\dfrac{5x^2+5x+2}{(x-2)^2(x+2)}$

2. $\log_7(x-2)+\log_7(x-4)=\log_7 15; x=$

A. No answer

B. -1 only

C. 7 only

D. -1 and 7 only

E. More than two answers

3. The center of the circle $x^2+y^2+8x-12y=7$ is

A. $(8, -12)$

B. $(-8, 12)$

C. $(4, -6)$

D. $(-4, 6)$

E. $(7, 7)$

4. The slope of a line perpendicular to $5x - 6y = 7$ is

A. $-\frac{6}{5}$

B. $\frac{6}{5}$

C. $\frac{5}{6}$

D. $-\frac{5}{6}$

E. $-\frac{7}{6}$

5. If $y = -4x^2 + 24x + 10$, the vertex is

A. (0, 10)

B. (1, 30)

C. (2, 42)

D. (3, 46)

E. (4, 48)

6. $\frac{x}{x-5} \geq \frac{x}{x+5}$, then

A. $x > 5$ or $x < -5$

B. $x > 5$ or $-5 < x < 0$

C. $x > 5$ or $-5 < x \leq 0$

D. $x \geq 5$ or $-5 \leq x \leq 0$

E. This is true for all real numbers.

7. A plane goes west from Chicago. A second plane going 100 mph faster goes east. After 3 hours the planes are 3,000 miles apart. The faster plane's speed in mph is

A. 450

B. 475

C. 500

D. 525

E. 550

8. If $\frac{1}{a} + \frac{1}{b} = \frac{1}{x}$, $x =$

A. $a + b$

B. $\frac{ab}{a+b}$

C. $\frac{a+b}{ab}$

D. $\frac{a-b}{ab}$

E. $\frac{b-a}{ab}$

9. $\sin 2x = 2 \sin x \cos x$ and x is in the second quadrant. If $\sin x = \frac{3}{5}$, then $\sin 2x =$

A. $\frac{6}{5}$

B. $\frac{24}{25}$

C. $-\frac{24}{25}$

D. $\frac{12}{13}$

E. $-\frac{12}{13}$

10. $\begin{vmatrix} 2 & 4 & 0 \\ -3 & 5 & 0 \\ 7 & 8 & -2 \end{vmatrix} =$

A. -44

B. 0

C. 44

D. 144

E. 288

11. How many different ways can six people sit in a circle?

A. 24

B. 120

C. 720

D. 5040

E. 40,320

12. A jar contains 6 red balls and 4 blue balls. Two balls are drawn with replacement. The probability that both balls are blue is

A. 0.16

B. 0.36

C. 0.4

D. 0.8

E. 1

13. If $0 \leq x < 360°$ and $\cos x + 2 \sin x \cos x = 0$, $x =$

A. 0°, 180°, 30°, 150°

B. 0°, 180°, 210°, 330°

C. 90°, 270°, 30°, 150°

D. 90°, 270°, 210°, 330°

E. 60°, 300°, 0°, 180°

14. The graph of $x^2 + 1 = y^2$ is a(n)

A. circle

B. parabola

C. ellipse

D. hyperbola

E. no graph

15. $y = -3 \cos\left(4x - \frac{\pi}{4}\right) + 11$, the left-right shift is

A. π

B. $-\pi$

C. $-\frac{\pi}{4}$

D. $-\frac{\pi}{16}$

E. $\frac{\pi}{16}$

16. If $f(x) = \sqrt{3x + 1}$, then $f^{-1}(x) =$

A. $\dfrac{1}{\sqrt{3x + 1}}$

B. $\sqrt{3x - 1}$

C. $\dfrac{x^2 + 1}{3}$

D. $\dfrac{x^2 - 1}{3}$

E. $x^2 + \dfrac{1}{3}$

17. If $2^n + 2^n + 2^n + 2^n = 2^{100}$; $n =$

A. $\sqrt{10}$

B. 10

C. 25

D. 50

E. 98

18. A 25% solution of alcohol is mixed with a 50% solution to give 20 ounces of a 40% solution. How many ounces of 25% solution are used?

A. 4

B. 8

C. 10

D. 12

E. 16

19. $\log_{\frac{1}{8}}\left(\frac{1}{4}\right) =$

A. $\frac{1}{32}$

B. $\frac{1}{16}$

C. $\frac{1}{8}$

D. $\frac{1}{2}$

E. $\frac{2}{3}$

20. $\cot x \sin x =$

A. $\cos x$

B. $\tan x$

C. $\sec x$

D. $\csc x$

E. $\cot x$

College-Level Mathematics: Test 3 Answers

1. **C:** $\frac{(3x+2)(x-2)+(2x+3)(x+2)}{(x-2)(x-2)(x+2)}$.

2. **C:** $x > 4$ and $\log_7(x-2)(x-4) = \log_7 15$; so $(x-2)(x-4) = x^2 - 6x + 8 = 15$; and $x^2 - 6x - 7 = (x-7)(x+1) = 0$; since $x > 4$, only C is the answer.

3. **D:** $x^2 + 8x + 16 + y^2 - 12y + 36 = (x+4)^2 + (y-6)^2 = 7 + 16 + 36 = 59$

4. **A:** Solving for y, we get $y = \frac{5}{6}x - \frac{7}{6}$; so $m = \frac{5}{6}$; the slope of a perpendicular line will have a slope that is the negative reciprocal, or $-\frac{6}{5}$.

5. **D:** $x = \frac{-b}{2a} = \frac{-24}{2(-4)} = 3$. We don't have to find y since D is the only point that has $x = 3$.

6. **C:** We cannot cross multiply; $\frac{x}{x-5} - \frac{x}{x+5} = \frac{x(x+5)-x(x-5)}{(x-5)(x+5)} = \frac{10x}{(x-5)(x+5)} \geq 0$; the bottom of the fraction can't be 0, but the top can.

7. **E:** $3x + 3(x + 100) = 3000$; $x = 450$; $x + 100 = 550$.

8. **B:** LCD $= abx$; multiply the equation by the LCD and we get $bx + ax = (a + b)x = ab$; so $x = \dfrac{ab}{(a+b)}$

9. **C:** In the second quadrant, if $\sin x = \dfrac{3}{5}$; then $r = 5$ and $y = 3$; then it is a $3 - 4 - 5$ right triangle, and $x = -4$; $\cos x = \dfrac{-4}{5}$ $2\left(\dfrac{3}{5}\right)\left(-\dfrac{4}{5}\right) = -\dfrac{24}{25}$.

10. **A:** $(2)(5)(-2) + 4(0)(7) + (0)(-3)(8) - 7(5)(0) - 8(0)(2) - (-2)(-3)(4)$

11. **B:** $5! = 120$.

12. **A:** $(.4)(.4) = 0.16$.

13. **D:** $\cos x(1 + 2\sin x) = 0$; $\cos x = 0$; so $x = 90^\circ$ and 270°; $\sin x = -\dfrac{1}{2}$ so $x = 210^\circ$ and 330°

14. **D:** $y^2 - x^2 = 1$.

15. **E:** The answer $= 0$; $\left(\dfrac{1}{4}\right)\left(\dfrac{\pi}{4}\right) = \dfrac{\pi}{16}$.

16. **D:** Let $f(x) = y = \sqrt{3x + 1}$; $y^2 = 3x + 1$; $x = \dfrac{(y^2 - 1)}{3}$; $xo\, f^{-1}(x) = \dfrac{(x^2 - 1)}{3}$ with notation change.

17. **E:** $2^n + 2^n + 2^n + 2^n = 2^{100}$; $4\left(2^n\right) = 2^{100}$; $2^2\left(2^n\right) = 2^{100}$; $2^{2+n} = 2^{100}$; $2 + n = 100$; $n = 98$

18. **B:** $.25x + .50(20 - x) = .40(20)$; $x = 8$.

19. **E:** $(2^{-3})^x = 2^{-2}$; so $-3x = -2$ and $x = \dfrac{2}{3}$

20. **A:** $\dfrac{\cos x}{\sin x} \times \sin x = \cos x$.

College-Level Mathematics: Test 3 Answer Key

Question #	Answer Choice	Chapter	Subtopic to Review
1	C	5	Adding Algebraic Fractions
2	C	14	Logs
3	D	16	Circles
4	A	9	Lines
5	D	14	Parabolas
6	C	6	Quadratic Inequalities
7	E	7	Distance Word Problem
8	B	6	Linear Equations
9	C	15	Trig
10	A	16	Determinants
11	B	16	Permutations
12	A	16	Probabilities
13	D	15	Trig Equations
14	D	16	Conic Sections
15	E	15	Trig Graphs
16	D	14	Inverse Functions
17	E	3	Exponents
18	B	7	Mixture Word Problems
19	E	14	Logs
20	A	15	Trig

College-Level Mathematics: Test 4

1. Solve for x: $4^{2x+3} = 8^{3x+2}$

A. -2

B. -1

C. 0

D. 1

E. 2

2. $F(x) = 2x + 5$ and $G(x) = \dfrac{x-5}{2}$; $F(G(x)) =$

A. x

B. $\dfrac{2x^2 - 5x - 25}{2}$

C. $\dfrac{2(2x+5)}{x-5}$

D. $\dfrac{2x^2 - 25}{2}$

E. $\dfrac{3x}{2} + \dfrac{5}{2}$

3. If $\log 2 = .30$ and $\log 3 = .48$, $\log 360 =$

A. 1.56

B. 2.56

C. 15.60

D. 0.144

E. 144

4. $\dfrac{x^2 - 9}{(x-3)^3} \times \dfrac{x^3 + x^2}{x^2 + 4x + 3} =$

A. $\dfrac{x^2}{(x-3)^2}$

B. $\dfrac{x^5}{(x-3)^2(x+1)}$

C. $\dfrac{x^2}{(x-3)^2(x-1)}$

D. $\dfrac{x^3 + 2x^2 - 9}{(x-3)^2(x^2+4x+3)}$

E. $\dfrac{x^4}{(x-3)^2(x-1)}$

5. The number of points $3x^2 - 4y^2 = -42$ and $5x^2 + 3y^2 = 16$ have in common is

A. 0

B. 1

C. 2

D. 3

E. 4

6. $\sin\left(\cos^{-1}\dfrac{5}{7}\right) =$

A. $\dfrac{7}{5}$

B. $\dfrac{5}{\sqrt{74}}$

C. $\dfrac{7}{\sqrt{74}}$

D. $\dfrac{\sqrt{24}}{5}$

E. $\dfrac{\sqrt{24}}{7}$

7. Six sandwiches and 4 drinks cost \$29, and 2 sandwiches and 4 drinks cost \$15. The cost of one drink and one sandwich together is

A. \$4.50

B. \$5.00

C. \$5.50

D. \$6.00

E. \$6.50

8. One focus of $\frac{(x-3)^2}{9} - \frac{(y-5)^2}{16} = 1$ is

A. (3, 5)

B. (0, 5)

C. (3, 9)

D. (3, 4)

E. (8, 5)

9. $4 \log_{125} 5 =$

A. 625

B. 25

C. 4

D. $\frac{4}{3}$

E. $\frac{1}{625}$

10. $i^{1776} =$

A. -1

B. 0

C. 1

D. $-i$

E. i

11. $\sqrt[10]{x^{147}} =$

A. $x^{\frac{10}{147}}$

B. x^{137}

C. $x^{73}\sqrt[10]{x}$

D. $x^{14}\sqrt[10]{x^7}$

E. $x^{12}\sqrt[10]{x^3}$

12. $e^{\ln 5} + 5 \log 10 + \log_7 1 =$

A. 1

B. 5

C. 10

D. 11

E. 25

13. Solve for x: $(x - 6)^2 = 4x(2x - 5) - 7x(x - 2)$

A. -6

B. 6

C. ± 6

D. 0, 6

E. 0, -6

14. $(4xy^6z^9)^{\frac{3}{2}} =$

A. $4xy^9z^{\frac{27}{2}}$

B. $4xy^4z^6$

C. $8xy^9z^{\frac{27}{2}}$

D. $8x^{\frac{3}{2}}y^9z^{\frac{27}{2}}$

E. $8x^{\frac{3}{2}}y^4z^6$

15. $C(100,2) =$

A. 10,000

B. 9900

C. 5000

D. 4950

E. 2425

16. The positive geometric mean of 10 and 6 is

A. 7

B. 7.5

C. 8

D. $2\sqrt{15}$

E. $\sqrt{6^7}$

17. $\dfrac{x^{4m}x^{5m+6}}{x^p} =$

A. $x^{\frac{(20m^2+24m)}{p}}$

B. $x^{9m+6} - x^p$

C. $x^{9m+6} - p$

D. x^{9m+6-p}

E. $\dfrac{1}{x^{p-9m+6}}$

18. If $0 \leq x \leq 360°$, $7\sin^3 x - \sin x = 0$ has how many answers?

A. 3

B. 4

C. 5

D. 6

E. More than 6

19. $\dfrac{\sqrt{7}}{\sqrt{7}-\sqrt{6}} =$

A. $\dfrac{7-\sqrt{6}}{13}$

B. $\dfrac{7+\sqrt{42}}{13}$

C. $7+\sqrt{42}$

D. $7-\sqrt{42}$

E. 49

20. How many diagonals are in a pentagon?

A. 5

B. 7

C. 10

D. 15

E. 20

College-Level Mathematics: Test 4 Answers

1. C: $2^{2(2x+3)} = 2^{3(3x+2)}$; so $4x + 6 = 9x + 6$; $0 = 5x$, $x = 0$.

2. A: $2\left(\dfrac{[x-5]}{2}\right) + 5 = x - 5 + 5 = x$

3. B: $\log 360 = \log(10 \times 4 \times 9) = \log 10 + 2\log 2 + 2\log 3 = 1 + .60 + .96 = 2.56$.

4. A: $\dfrac{(x+3)(x-3)}{(x-3)(x-3)(x-3)} \times \dfrac{x^2(x+1)}{(x+1)(x+3)}$ cancel one $(x-3)$, $(x+3)$, and $(x+1)$ top and bottom.

5. A: $3(3x^2 - 4y^2) + 4(5x^2 + 3y^2) = 29x^2 = 3(-42) + 4(16) = -62$; if $29x^2 = -62$, there are no real solutions.

6. E: Letting $x = 5$ and $r = 7$, we get $y = \sqrt{24}$; $\sin = \dfrac{y}{r}$

7. C: $6s + 4d = 29$ and $2s + 4d = 15$; adding we get $8s + 8d = 44$; so $s + d = \dfrac{44}{8} = \$5.50$.

8. E: Where the asymptotes cross is (3, 5); $c^2 = 9 + 16 = 25$; so $c = 5$; foci are $(3 \pm 5, 5)$; E is the only answer that fits.

9. D: $\log_{125} 5^4 = x$; so $5^{3x} = 5^4$ and $3x = 4$.

10. C: $\frac{1776}{4}$ has a remainder of 0, and $i^0 = 1$.

11. D: 10 into 147 = 14 remainder 7; x^{14} on the outside; x^7 under the radical sign.

12. C: $e^{\ln x} = 5$; so the first term is 5; log10 (to the base 10) = 1; 1 times 5 = 5, the second term; $\log_b 1$ is always 0 to any base; so 5 + 5 + 0 = 10.

13. B: $x^2 - 12x + 36 = 8x^2 - 20x - 7x^2 + 14x$; $-12x + 36 = -6x$; $-6x = -36$.

14. D: $(\sqrt{4})^3 x^{\frac{3}{2}} y^{6\left(\frac{3}{2}\right)} z^{9\left(\frac{3}{2}\right)} = 8x^{\frac{3}{2}} y^9 z^{\frac{27}{2}}$

15. D: $\frac{100 \times 99}{2 \times 1} = 4950.$

16. D: $\sqrt{6 \times 10} = 2\sqrt{15} = \sqrt{4 \times 15}$

17. D: $x^{4m+5m+6-p}$ when you multiply, you add exponents; when you divide, you subtract exponents.

18. E: $\sin x\,(7 \sin^2 x - 1) = 0$. $\sin x = 0$ mean $x = 0$; 180° or 360°. If $7 \sin^2 x - 1 = 0$, then $x = \sin^{-1} \pm \sqrt{\frac{1}{7}}$; $\sin^{-1} \sqrt{\frac{1}{7}} \approx 22°$ or 158°; $\sin^{-1} - \sqrt{\frac{1}{7}} \approx 202°$ or 338°

19. C: Multiply top and bottom by $\sqrt{7} + \sqrt{6}$. $\frac{\sqrt{7}}{\sqrt{7} - \sqrt{6}} \left(\frac{\sqrt{7} + \sqrt{6}}{\sqrt{7} + \sqrt{6}}\right) = \left(\frac{7 + \sqrt{42}}{7 - 6}\right) = 7 + \sqrt{42}$

20. A: $d = \frac{n(n-3)}{2} = \frac{5(5-3)}{2} = \frac{5(2)}{2} = 5.$

College-Level Mathematics: Test 4 Answer Key

Question #	Answer Choice	Chapter	Subtopic to Review
1	C	3	Exponents
2	A	14	Composite Functions
3	B	14	Logs
4	A	5	Multiplying Algebraic Functions
5	A	8	Solving Two Non-linear Equations in Two Unknowns
6	E	15	Inverse Trig Functions
7	C	8	Word Problems Involving Two Equations in Two Unknowns
8	E	16	Hyperbolas
9	D	14	Logs
10	C	4	Complex Numbers
11	D	3	Fractional Exponents
12	C	14	Logs
13	B	6	Solving Linear Equations (Plus Algebraic Simplifications)
14	D	3	Exponents
15	D	16	Combinations
16	D	16	Geometric Progressions
17	D	3	Exponents
18	E	15	Trig Equations
19	C	4	Square Roots
20	A	11	Polygons

INDEX

A

Abscissa 131
Absolute Value
  Equalities 93
  Graphs 210
  Inequalities 94
Algebraic Division 82
Angles
  Acute 147
  Complementary 148
  Exterior 154
  Formed by Parallel Lines 148
  Obtuse 147
  Reflex 148
  Right 147
  Straight 148
  Supplementary 148
  Vertical 149
Area
  Circle 184
  Parallelogram 169
  Rectangle 173
  Rhombus 172
  Square 175
  Trapezoid 177
  Triangle 152
Arithmetic
  Decimals 13
  Fractions 17
  Percentages 25
  Rounding Off 15
Arrangements 252
Asymptotes 272, 280
Avoiding Duplicates 253
Axis (Axes) 131

B

Bisect Definition 188
Box 195

C

Circle
  Arc 184
  Center 143, 183
  Chord 184
  Circumference 183
  Diameter 184
  Equation of 143
  Point of Tangency 184
  Radius 143, 183
  Secant 184
  Semicircle 184
  Sector 184
  Tangent 184
Circumference Formula 184
Combinations 253
Cone 201
Conic Sections
  Circles 268
  Ellipses 269
  Hyperbolas 274, 276
  Parabolas 276
Conjugate 48, 56
Cosecant 227
Cotangent 227
Counting 251
Cube 196
Cube Root 48
Curve Sketching 279
Cylinder 197

D

Determinants 259
Diagonal
  Box 195
  Cube 196
  Number in a Polygon 179
  Rectangle 173
  Rhombus 172
  Square 175
  Trapezoid 177
Diameter 184
Difference 2
Distance Formula 133
Distributive Law 62
Dividend 82
Divisibility Rules 3
Divisor 2, 82

E

Equations
  Absolute Value 93
  First Degree 87
  Linear 87
  Quadratic 96, 98
  Square Roots 283
  Two or More Unknowns 121, 123
Exponential Growth and Decay 221
Exponents 35

F

Factors 2, 67
Factoring 67
FOIL 62
Fractions-Algebraic 76
Functions
  Composition 212
  Domain 205
  Inverse 213
  Map 205
  1 to 1 213
  Piecewise 206
  Range 205

I

Inequalities 1
  Absolute Value 95
  Linear 91
  Quadratic 102
Integers 1

L

Like Terms 61
Line
  Equation of 141
  Horizontal 134
  Parallel 141
  Perpendicular 141
  Point-Slope 138
  Slope-Intercept 139
  Standard Form 138
  Vertical 134
  x-intercept 138
  y-intercept 138
Logarithms 215

M

Magnitude 55
Matrix (Matrices) 258
Mean 6
Measurements 116
Median 6
Midpoint Definition 188
Midpoint Formula 133
Mode 6
Multiples 2
  LCM 20

N

Numbers
  Complex 54
  Composites 3
  Divisors 2
  Even 2
  Factors 2
  Fibonacci 283
  Imaginary 54
  Integers 1
  Multiple 2
  Odd 2
  Operations 2
  Perfect 3
  Prime 2
  Rational 18
  Whole 1

O

Order of Operations 4
Ordered Pairs 131
Ordinate 131
Origin 132

P

Parabolas 220, 276
Parallelogram 169
Perimeter
  Parallelogram 169
  Rectangle 173
  Rhombus 172
  Square 175
  Trapezoid 177
  Triangle 152
Permutations 252
Point 131
Polygons 179
Polynomials-Sketch Intercepts 279
Probability 255
Product 2
Progressions
  Arithmetic 264
  Geometric-Finite 266
  Geometric-Infinite 267
Pyramid 198
Pythagorean Theorem 159
Pythagorean Triples 161

Q

Quadrant 132
Quadratic Formula 98
Quadrilaterals 169
Quotient 2, 82

R

Radian 225
Radius 143, 183
Range 6, 205
Ratio of Similarity 165
Rational Functions, Sketch
  Asymptotes 280
  Intercepts 279
Rectangle 173
Rhombus 172

S

Scientific Notation 43
Secant 184, 227
Sectors 185
Sequences 282
Sets 256
Sine 227
Slant Height 201
Slope 134
Sphere 200
Square 175
Square Root 47

Statistics
Mean 6
Median 6
Mode 6
Range 6
Sum 2
Surface Area
Box 195
Cone 201
Cube 196
Cylinder 197
Sphere 200

T

Tangent 184, 227
Trapezoid 177
Base Angles 177
Diagonals 177
Isosceles 177
Translations 261
Triangles
Area 152
Acute 153
Altitude (Height) 152
Base 152
Base Angles 153
Bisector 152
Equiangular 153
Equilateral 153
Exterior 154
45° – 45° – 90° 161
Hypotenuse 153
Isosceles 153
Leg 153
Median 153
Obtuse 154
Perimeter 152
Right 153
Scalene 153
Similar 165
Sum of Angles 152
30° – 60° – 90° 161
Trigonometry
Equations 240
Identities 239
Inverse 237
Law of Cosines 245
Law of Sines 245
Radians 225
Ratios 226
Right Angle 243
Sketching Sines and Cosines 232

U

Unlike Terms 61

V

Volumes
Box 195
Cone 201
Cube 196
Cylinder 197
Sphere 200

W

Whole Numbers 1
Word Problems
Basics 107
Distance 111
Everyday 109
Interest 117
Ratios 110
Speed 111

NOTES

NOTES

NOTES

NOTES

NOTES

NOTES

NOTES

NOTES

NOTES

NOTES

NOTES

REA's Test Preps

The Best in Test Preparation

- *REA Test Preps are **far more** comprehensive than any other test preparation series*
- *Each book contains full-length practice tests based on the most recent exams*
- ***Every** type of question likely to be given on the exams is included*
- *Answers are accompanied by **full** and **detailed** explanations*

Some of our titles include:

Advanced Placement Exams (APs)
Art History
Biology
Calculus AB & BC
Chemistry
Economics
English Language & Composition
English Literature & Composition
Environmental Science
European History
French Language
Government & Politics
Human Geography
Physics
Psychology
Spanish Language
Statistics
United States History
World History

SAT Subject Tests
Biology E/M
Chemistry
Latin
Literature
Mathematics Level 2
Spanish
United States History
World History

ACCUPLACER

ACT - ACT Assessment

ASVAB - Armed Services Vocational Aptitude Battery

College-Level Examination Program (CLEP)
American Government
American Literature
Biology
Calculus
Chemistry
College Algebra
College Composition
College Mathematics
Financial Accounting
History of the United States I + II
Human Growth and Development
Introduction to Educational Psychology
Introductory Business Law
Introductory Psychology
Introductory Sociology
Natural Sciences
Principles of Macroeconomics
Principles of Management
Principles of Marketing
Principles of Microeconomics
Spanish Language
Western Civilization I + II

CBEST - California Basic Educational Skills Test

CDL - Commercial Driver License Exam

COOP, HSPT & TACHS - Catholic High School Admission Tests

EMT

FTCE - Florida Teacher Certification Examinations

GED® Test

GRE Exams

LSAT - Law School Admission Test

MAT - Miller Analogies Test

MTEL - Massachusetts Tests for Educator Licensure

NCLEX-PN

NYSTCE - New York State Teacher Certification Examinations

PANCE - Physician Assistant National Certifying Exam

PCAT - Pharmacy College Admission Test

PTCE - Pharmacy Technician Certification Exam

PRAXIS

TEAS

TExES - Texas Examinations of Educator Standards

TOEFL - Test of English as a Foreign Language

For our complete title list, visit www.rea.com

Research & Education Association

REA's Study Guides

Review Books, Refreshers, and Comprehensive References

Problem Solvers®
Presenting an answer to the pressing need for easy-to-understand and up-to-date study guides detailing the wide world of mathematics and science.

High School Tutors®
In-depth guides that cover the length and breadth of the science and math subjects taught in high schools nationwide.

Essentials®
An insightful series of useful, practical, and informative references comprehensively covering popular subjects.

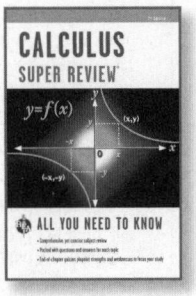

Super Reviews®
Don't miss a thing! Review it all thoroughly with this series of complete subject references at an affordable price.

Interactive Flashcard Books®
Flip through these essential, interactive study aids that go far beyond ordinary flashcards.

ESL
Easy-to-use reference for anyone who is learning the English language and wants to improve listening, speaking and writing skills.

For our complete title list, visit www.rea.com

Research & Education Association